세상에서 가장 쉬운 과학 수업

양자전기역학

세상에서 가장 쉬운 과학 수업
양자전기역학

ⓒ 정완상, 2025

초판 1쇄 인쇄 2025년 3월 10일
초판 1쇄 발행 2025년 3월 25일

지은이 정완상
펴낸이 이성림
펴낸곳 성림북스

책임편집 최윤정
디자인 쏘울기획

출판등록 2014년 9월 3일 제25100-2014-000054호
주소 서울시 은평구 연서로3길 12-8, 502
대표전화 02-356-5762
팩스 02-356-5769
이메일 sunglimonebooks@naver.com

ISBN 979-11-93357-46-0   03400

* 책값은 뒤표지에 있습니다.
* 이 책의 판권은 지은이와 성림북스에 있습니다.
* 이 책의 내용 전부 또는 일부를 재사용하려면 반드시 양측의 서면 동의를 받아야 합니다.

노벨상 수상자들의 **오리지널 논문으로 배우는** 과학

세상에서 가장 쉬운 과학 수업

양자 전기역학

정완상 지음

선형대수학의 역사부터 파인먼 다이어그램까지
양자전기역학의 창시자와 그의 혁명적 논문을 만나다

성림원북스

CONTENTS

과학을 처음 공부할 때 이런 책이 있었다면 얼마나 좋았을까   007
천재 과학자들의 오리지널 논문을 이해하게 되길 바라며   010
양자전기역학의 창시자 파인먼_글래쇼 박사 깜짝 인터뷰   013

첫 번째 만남
선형대수학의 역사 / 017

선형연립방정식_고대의 연구 기록   018
크라메르의 공식과 가우스 소거법_연립방정식의 새로운 해법 발견   022
행렬의 등장_직사각형 모양으로 수 배열하기   033
벡터의 탄생_오래 누워 있는 습관 덕에   042

그라스만의 벡터 공간_체를 이루는 조건　　　　　　　　　　056

두 번째 만남
디랙의 양자역학 / 061

디랙 브라켓의 탄생_벡터를 표현하는 또 다른 기호　　　　　062
코시 적분_새로운 적분 공식　　　　　　　　　　　　　　　071
디랙 델타 함수_디랙이 만든 함수의 성질　　　　　　　　　078
하이젠베르크-보른-요르단-슈뢰딩거의 양자역학_양자역학의 탄생 과정　　083
양자역학의 위치 상태와 운동량 상태_전자의 위치와 운동량　　091

세 번째 만남
파인먼의 경로 적분 / 109

파인먼의 생애_수학 천재의 사랑과 도전　　　　　　　　　110
파인먼의 아이디어_양자역학을 새롭게 이해하는 이론　　　119
하위헌스의 원리_과거의 정보로부터 현재가 결정되다　　　123
경로 적분_전자를 발견할 확률 구하기　　　　　　　　　　129
경로 적분을 구하는 방법_모든 가능한 경로에 대해　　　　137
전파인자의 계산_전자가 힘을 받지 않는 경우와 힘을 받는 경우　　145

네 번째 만남

양자전기역학 / 157

양자장론을 만든 물리학자들_포크, 요르단, 위그너	158
양자장론_입자의 생성과 소멸에 대한 이론	166
양자전기역학의 창시자_슈윙거와 도모나가 신이치로	177
양자전기역학_파인먼 다이어그램을 이용하여	182

만남에 덧붙여 / 193

A New Notation For Quantum Mechanics_디랙 논문 영문본	194
Space-Time Approach to Non-Relativistic Quantum Mechanics_파인먼 1 논문 영문본	197
Space-Time Approach to Quantum Electrodynamics_파인먼 2 논문 영문본	218
위대한 논문과의 만남을 마무리하며	239
이 책을 위해 참고한 논문들	242
수식에 사용하는 그리스 문자	244
노벨 물리학상 수상자들을 소개합니다	245

과학을 처음 공부할 때 이런 책이 있었다면 얼마나 좋았을까

남순건(경희대학교 이과대학 물리학과 교수 및 전 부총장)

21세기를 20여 년 지낸 이 시점에서 세상은 또 엄청난 변화를 맞이하리라는 생각이 듭니다. 100년 전 찾아왔던 양자역학은 반도체, 레이저 등을 위시하여 나노의 세계를 인간이 이해하도록 하였고, 120년 전 아인슈타인에 의해 밝혀진 시간과 공간의 원리인 상대성이론은 이 광대한 우주가 어떤 모습으로 만들어져 왔고 앞으로 어떻게 진화할 것인가를 알게 해주었습니다. 게다가 우리가 사용하는 모든 에너지의 근원인 태양에너지를 핵융합을 통해 지구상에서 구현하려는 노력도 상대론에서 나오는 그 유명한 질량-에너지 공식이 있기에 조만간 성과가 있을 것이라 기대하게 되었습니다.

앞으로 올 22세기에는 어떤 세상이 펼쳐질지 매우 궁금합니다. 특히 인공지능의 한계가 과연 무엇일지, 또한 생로병사와 관련된 생명의 신비가 밝혀져 인간 사회를 어떻게 바꿀지, 우주에서는 어떤 신비로움이 기다리고 있는지, 우리는 불확실성이 가득한 미래를 향해 달려가고 있습니다. 이러한 불확실한 미래를 들여다보는 유리구슬 역할을 하는 것이 바로 과학적 원리들입니다.

지난 백여 년간 과학에서의 엄청난 발전들은 세상의 원리를 꿰뚫어보았던 과학자들의 통찰을 통해 우리에게 알려졌습니다. 이런 과학 발전을 가능하게 한 영웅들의 생생한 숨결을 직접 느끼려면 그들이 썼던 논문들을 경험해보는 것이 좋습니다. 그런데 어느 순간 일반인과 과학을 배우는 학생들은 물론, 그 분야에서 연구를 하는 과학자들마저 이런 숨결을 직접 경험하지 못하고 이를 소화해서 정리해놓은 교과서나 서적들을 통해서만 접하고 있습니다. 창의적인 생각의 흐름을 직접 접하는 것은 그런 생각을 했던 과학자들의 어깨 위에서 더 멀리 바라보고 새로운 발견을 하고자 하는 사람들에게 매우 중요합니다.

저자인 정완상 교수가 새로운 시도로써 이러한 숨결을 우리에게 전해주려 한다고 하여 그의 30년 지기인 저는 매우 기뻤습니다. 그는 대학원생 때부터 당시 혁명기를 지나면서 폭발적인 발전을 하고 있던 끈 이론을 위시한 이론물리학 분야에서 가장 많은 논문을 썼던 사람입니다. 그리고 그러한 에너지가 일반인들과 과학도들을 위한 그의 수많은 서적을 통해 이미 잘 알려져 있습니다. 저자는 이번에 아주 새로운 시도를 하고 있고 이는 어쩌면 우리에게 꼭 필요했던 것일 수 있습니다. 대화체로 과학의 역사와 배경을 매우 재미있게 설명하고, 그 배경 뒤에 나왔던 과학 영웅들의 오리지널 논문들을 풀어간 것입니다. 과학사를 들려주는 책들은 많이 있으나 이처럼 일반인과 과학도의 입장에서 질문하고 이해하는 생각의 흐름을 따라 설명한 책은 없습니다. 게다가 이런 준비를 마친 후에 아인슈타인 같은 영웅들의 논문을 원래의 방식과 표기를 통해 설명하는 부분은 오랫동안 과학을 연구해온 과학자에게도 도움을 줍니다.

이 책을 읽는 독자들은 복 받은 분들일 것이 분명합니다. 제가 과학을 처음 공부할 때 이런 책이 있었다면 얼마나 좋았을까 하는 생각이 듭니다. 정완상 교수는 이제 새로운 형태의 시리즈를 시작하고 있습니다. 독보적인 필력과 독자에게 다가가는 그의 친밀성이 이 시리즈를 통해 재미있고 유익한 과학으로 전해지길 바랍니다. 그리하여 과학을 멀리하는 21세기의 한국인들에게 과학에 대한 붐이 일기를 기대합니다. 22세기를 준비해야 하는 우리에게는 이런 붐이 꼭 있어야 하기 때문입니다.

천재 과학자들의 오리지널 논문을
이해하게 되길 바라며

　사람들은 과학 특히 물리학 하면 너무 어렵다고 생각하지요. 제가 외국인들을 만나서 얘기할 때마다 신선하게 느끼는 점이 있습니다. 그들은 고등학교까지 과학을 너무 재미있게 배웠다고 하더군요. 그래서인지 과학에 대해 상당한 지식을 가진 사람들이 많았습니다. 그 덕분에 노벨 과학상도 많이 나오는 게 아닐까 생각해요. 우리나라는 노벨 과학상 수상자가 한 명도 없습니다. 이제 청소년과 일반 독자의 과학 수준을 높여 노벨 과학상 수상자가 매년 나오는 나라가 되게 하고 싶다는 게 제 소망입니다.

　그동안 양자역학과 상대성이론에 관한 책은 전 세계적으로 헤아릴 수 없을 정도로 많이 나왔고 앞으로도 계속 나오겠지요. 대부분의 책은 수식을 피하고 관련된 역사 이야기들 중심으로 쓰여 있어요. 제가 보기에는 독자를 고려하여 수식을 너무 배제하는 것 같았습니다. 이제는 독자들의 수준도 많이 높아졌으니 수식을 피하지 말고 천재 과학자들의 오리지널 논문을 이해하길 바랐습니다. 그래서 앞으로 도래할 양자(量子, quantum)와 상대성 우주의 시대를 멋지게 맞이하도록 도우리라는 생각에서 이 기획을 하게 된 것입니다.

원고를 쓰기 위해 논문을 읽고 또 읽으면서 어떻게 이 어려운 논문을 독자들에게 알기 쉽게 설명할까 고민했습니다. 여기서 제가 설정한 독자는 고등학교 정도의 수식을 이해하는 청소년과 일반 독자입니다. 물론 이 시리즈의 논문에 그 수준을 넘어서는 내용도 나오지만 고등학교 수학만 알면 이해할 수 있도록 설명했습니다. 이 책을 읽으며 천재 과학자들의 오리지널 논문을 얼마나 이해할지는 독자들에 따라 다를 거라 생각합니다. 책을 다 읽고 100% 혹은 70%를 이해하거나 30% 미만으로 이해하는 독자도 있을 것입니다. 저의 생각으로는 이 책의 30% 이상 이해한다면 그 사람은 대단하다고 봅니다.

이 책에서는 양자전기역학의 창시자 중 한 명인 파인먼의 논문을 다루었습니다. 여기에 쓰이는 브라켓 기호를 설명하고자 디랙의 오래된 논문 내용을 넣었습니다. 이것을 이해하기 위해서는 선형대수학이라는 수학의 한 분야가 필요한데 1장에서 선형대수학의 역사를 소개했고, 2장에서는 디랙의 새로운 벡터 기호인 브라켓을 자세히 설명했습니다.

파인먼은 물리학을 즐겁게 연구한 과학자입니다. 그가 대학 시절에 양자역학을 공부하면서 떠올린 경로 적분 아이디어는 정말로 놀라울 따름입니다. 교과서가 아닌 자기 생각으로 양자역학을 공부한 파인먼과 학점을 따기 위해 양자역학을 공부하는 우리나라 학생들의 모습을 비교하면 너무 큰 차이가 느껴집니다. 젊은 과학자들이 파인먼처럼 물리를 생각하고 물리를 즐기기 바랍니다. 그것이 세상을 깜짝 놀라게 할 일을 만들 수 있으니까요.

마지막으로는 파인먼의 양자전기역학 논문의 내용을 수식을 줄여서 다루어 보았습니다. 파인먼 다이어그램을 통해 여러분이 대학원 수준의 양자전기역학 내용을 쉽게 이해할 수 있으리라 생각합니다.

〈노벨상 수상자들의 오리지널 논문으로 배우는 과학〉 시리즈는 많은 이에게 도움을 줄 수 있다고 생각합니다. 과학자가 꿈인 학생과 그의 부모, 어릴 때부터 수학과 과학을 사랑했던 어른, 양자역학과 상대성이론을 좀 더 알고 싶은 사람, 아이들에게 위대한 논문을 소개하려는 과학 선생님, 반도체나 양자 암호 시스템, 우주 항공 계통 등의 일에 종사하는 직장인, 〈인터스텔라〉를 능가하는 SF 영화를 만들고 싶어 하는 영화 제작자나 웹툰 작가 등 많은 사람들에게 이 시리즈를 추천합니다.

진주에서 정완상 교수

양자전기역학의 창시자 파인먼
_ 글래쇼 박사 깜짝 인터뷰

빛과 물질의 상호작용을 다루다

기자　오늘은 파인먼의 1949년 양자전기역학 논문에 대해 글래쇼 박사와 인터뷰를 진행하겠습니다. 글래쇼 박사는 약전자기 통일 이론으로 1979년 노벨 물리학상을 수상한 분이지요. 글래쇼 박사님, 나와 주셔서 감사합니다.

글래쇼　제가 제일 존경하는 과학자인 파인먼의 논문에 관한 내용이라 만사를 제치고 달려왔습니다.

기자　파인먼은 양자전기역학의 창시자라고 합니다. 양자전기역학이란 무엇인가요?

글래쇼　양자전기역학의 창시자는 파인먼과 슈윙거, 도모나가 신이치로 이렇게 세 사람입니다. 그중에서 대중에게 가장 많이 알려진 사람은 파인먼입니다. 그는 양자전기역학의 복잡한 계산식을 파인먼 다이어그램이라는 아주 예쁜 그림으로 나타냈지요. 양자전기역학은 이름 그대로 전기역학을 양자화한 이론입니다. 이 이론이 나오기 전에 양자장론이 먼저 등장하죠.

기자　양자장론은 뭐죠?

글래쇼　양자역학은 위치와 운동량의 불확정성을 동시에 0으로 만들

수 없다는 이론이고, 양자장론은 입자의 생성과 소멸에 대한 이론입니다. 이 이론을 토대로 빛과 물질의 상호작용을 다루는 것이 양자전기역학이지요.

기자　그렇군요.

파인먼의 경로 적분

기자　파인먼은 양자역학을 새로운 방법으로 설명했다고 하는데요. 어떤 내용이죠?

글래쇼　경로 적분을 말씀하시는군요. 파인먼은 하위헌스의 원리처럼 과거의 정보가 모여서 현재의 정보를 만드는 과정과 비슷하게 양자역학을 묘사할 수 있다고 생각했어요. 그러니까 시간에 따라 위치가 달라지는데 과거의 입자 위치로부터 미래의 입자 위치를 가능한 모든 경로를 고려한 적분을 이용해 나타낼 수 있다는 겁니다.

기자　아주 멋진 아이디어네요.

파인먼의 1949년 논문 개요

기자　파인먼의 1949년 논문에는 어떤 내용이 담겨 있나요?

글래쇼　파인먼은 양자역학에서 도입한 경로 적분을 양자장론에도 적용했습니다. 그리고 광자와 전자의 상호작용을 다루기 위해 여러 가지 파인먼 다이어그램을 고려했지요. 즉, 모든 가능한 파인먼 다이어그램을 통해 확률진폭을 계산하는 것입니다. 아주 복잡한 과정이지만 파인먼은 간단한 그림으로 많은 사람이 쉽게 이해할 수 있게 해주었지요.

기자   실험적으로 관측한 예가 있나요?

글래쇼   물론입니다. 전자의 자기모멘트를 양자역학적으로 계산한 결과와 실험값 사이의 오차가 있었는데 양자전기역학을 통해 이 오차가 거의 0이 되었습니다. 즉, 양자전기역학이 옳다는 것이 증명된 거죠.

기자   좀 더 자세히 알아보고 싶군요.

파인먼의 1949년 논문이 일으킨 파장

기자   파인먼의 1949년 논문은 어떤 변화를 가지고 왔나요?

글래쇼   파인먼의 경로 적분과 파인먼 다이어그램 덕분에 입자 물리학자들은 양자장론과 양자전기역학을 쉽게 이해하게 되었습니다. 이 방법은 뒤에 약전자기 통일 이론이나 쿼크 이론을 다룰 때도 쓰였지요. 즉, 양자전기역학의 혁명 이후에 소립자 물리학은 급진적으로 발전했고 결국 약전자기 통일 이론까지 성공을 거두었습니다.

기자   엄청나게 중요한 역할을 했군요. 지금까지 파인먼의 양자전기역학 논문에 대해 글래쇼 박사의 이야기를 들어 보았습니다.

첫 번째 만남

선형대수학의 역사

선형연립방정식 _ 고대의 연구 기록

정교수   우리는 먼저 '선형(linearity)'과 '선형연립방정식'의 역사를 알아볼 거야. 예를 들어 다음 함수를 봐.

$f(x) = x$

이 함수는 다음과 같은 두 가지 성질을 만족해.

$f(kx) = kf(x)$ (k는 어떤 수)

$f(x+y) = f(x) + f(y)$

위 두 조건을 만족할 때 함수 $f(x)$는 선형이라고 말해. 만일 선형이 아니면 비선형이라고 하지.

물리군 $f(x) = x^2$은 선형이 아니네요. $f(kx) = k^2 x^2$이고 $kf(x) = kx^2$으로 서로 다르니까요.

정교수   맞아. $f(x) = x^2, f(x) = x^3, f(x) = x^4, \cdots$ 은 모두 비선형이야. 이제 선형연립방정식을 생각해 볼게. 다음과 같이 미지수 두 개로 이루어진 두 개의 방정식을 선형연립방정식이라고 불러.

$x + y = 3$

$x - y = 7$

물리군   왜 선형연립방정식이라고 부르죠?

정교수   선형항인 x, y로만 이루어져 있기 때문이야. 그러니까 다음과 같은 연립방정식은 선형연립방정식이 아니야.

$$x^2 + y^2 = 5$$

$$x + y = 3$$

물리군 x^2이나 y^2이 선형이 아니기 때문이군요.

정교수   그렇지.

물리군   선형연립방정식은 누가 처음 생각했나요?

정교수   너무 오래전에 알려졌기 때문에 처음 이 문제를 다룬 사람에 대해서는 전해지지 않았어. 고대 바빌로니아인들이 기원전 4세기경에 선형연립방정식으로 이어지는 문제들을 연구했다는 기록이 있지. 그 후에 중국 수학자들이 선형연립방정식을 푸는 문제를 다루었어. 선형연립방정식은 연립일차방정식이라고도 부른다네.

    중국은 기원전 1600년경에 세워진 상(商)나라로부터 시작된다. 상나라는 기원전 1046년경까지 존재한 왕조이다. 상나라의 수도가 은(殷)이기 때문에 은나라로 부르기도 한다. 그 뒤를 이은 두 번째 왕조는 주(周)나라이다. 주나라는 기원전 1046년부터 기원전 256년까지 790년간 이어져 중국에서 가장 오랜 기간 존속한 나라이다.

    중국이 언제부터 수학 연구를 시작했는지는 정확히 알 수 없지만 중국 수학 고전 중 가장 오래된 책인 《주비산경(周髀算經)》에 수학과 천문학에 대한 내용이 기록되어 있다. 여기서 주(周)는 주나라를 뜻하고 비(髀)는 해시계의 바늘을 뜻한다. 《주비산경》이 언제 쓰인 책인지 자세히 알려져 있지 않지만, 학자들은 이 책이 기원전 300년경에 쓰였을 것으로 추정한다.

《주비산경》은 천문과 역법 및 기하학을 다루었는데 가장 대표적인 것은 피타고라스 정리이다.

'직사각형의 절반에서 가로의 길이가 3이고 세로의 길이가 4라면 대각선의 길이는 5이다. (故折矩, 以爲句廣三, 股修四, 徑隅五。)'

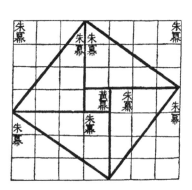

《주비산경》과 더불어 유명한 중국의 수학 고전은 《구장산술》이다. 이 책도 연대가 정확하지는 않지만 기원전 250년경에 쓰인 것으로 추정된다. 《구장산술》은 제목과 같이 9개의 장으로 구성되어 있는데 각 장에서 다루는 내용은 다음과 같다.

1장 다양한 평면도형의 넓이
2장 단순한 비례배분 문제
3장 복잡한 비례배분 문제
4장 제곱근과 세제곱근
5장 다양한 입체도형의 부피
6장 여러 가지 비례 문제
7장 일차방정식의 해
8장 연립일차방정식의 해
9장 피타고라스 정리

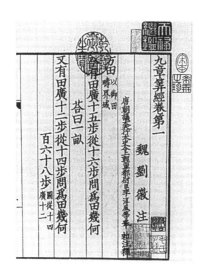

크라메르의 공식과 가우스 소거법 _ 연립방정식의 새로운 해법 발견

정교수   선형대수학은 벡터 공간, 벡터, 선형변환, 행렬, 선형연립방정식 등을 연구하는 대수학의 한 분야를 말해. 그 역사를 자세히 살펴보기로 하지.

선형대수학의 기초인 행렬과 행렬식에 대한 연구는 모두 선형연립방정식의 연구에서 비롯되었다. 흥미로운 것은 행렬의 개념이 행렬식의 개념보다 훨씬 나중에 소개되었다는 점이다.

행렬식의 개념을 처음 도입한 사람은 일본의 세키 다카카즈이다.

세키 다카카즈(関孝和, 1642~1708)

세키 다카카즈는 일본의 수학자이자 에도 시대(1603~1868)의 작가였다. 그는 고슈 한의 신하인 우치야마 가문에서 태어나, 쇼군의 신하인 세키 가문의 양자가 되었다. 고슈 한에 있는 동안 그는 고용주의 토지에 대한 신뢰할 만한 지도를 제작하는 측량 프로젝트에 참여했다. 그는 당시 일본에서 사용된 정확도가 떨어지는 달력을 대체하기 위해 13세기 중국 달력을 연구하는 데 수년을 보냈다. 또한 새로운 대수 표기법을 창안했으며 스위스의 베르누이보다 앞서 베르누이수를 발견했다.

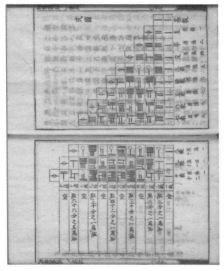

세키 다카카즈가 발견한 베르누이수

　한편 유럽에서 선형연립방정식의 새로운 풀이법을 처음 알아낸 사람은 크라메르이다.

크라메르(Gabriel Cramer, 1704~1752)

크라메르는 스위스 제네바에서 태어났다. 어릴 때부터 수학에 재능을 보인 그는 18세에 제네바 대학에서 박사 학위를 받았고 20세에는 제네바 대학 수학 교수가 되었다.

40대에는 대수곡선에 관한 논문(1750)을 발표했고, 행성의 구형에 대한 물리적 원인(1730)과 뉴턴의 입방 곡선 처리에 관한 논문(1746)을 썼다.

1750년에 크라메르는 행렬식을 도입해 연립일차방정식의 해의 공식을 찾아냈는데 이것을 크라메르의 공식이라고 부른다. 이 내용은 그의 책 《대수곡선의 해석에 관한 연구》에 수록되어 있다.

크라메르는 1730년대 후반부터

유럽 전역을 광범위하게 여행했는데, 이는 그의 수학 연구에 큰 영향을 미쳤다. 1752년 그는 건강을 회복하기 위해 프랑스 남부를 여행하던 중 바뇰쉬르세즈에서 사망했다.

물리군 크라메르의 공식이 뭐예요?
정교수 다음 연립방정식을 가지고 설명해 볼게.

$$ax + by = P \qquad (1\text{-}2\text{-}1)$$

$$cx + dy = Q \qquad (1\text{-}2\text{-}2)$$

식 (1-2-1)에 c를 곱하고 식 (1-2-2)에 a를 곱해서 빼면

$$(ad - bc)y = aQ - cP \qquad (1\text{-}2\text{-}3)$$

가 된다. 이때 $ad - bc$가 0인지 0이 아닌지가 아주 중요하다. 이것을 주어진 연립방정식의 계수들에 대한 행렬식(determinant)[1]이라 하고

$$\begin{vmatrix} a & b \\ c & d \end{vmatrix} = ad - bc \qquad (1\text{-}2\text{-}4)$$

로 정의한다.

---

[1] 행렬을 나타내는 영어는 matrix이고 행렬식을 나타내는 영어는 determinant이다. determinant는 원래 근이 생기는가 안 생기는가를 결정하는 역할을 하므로 '결정식'이라고 번역하는 것이 옳아 보인다. 하지만 우리보다 수학 번역을 먼저 한 일본이 이것을 행렬식으로 번역했고, 이 내용이 일제강점기 때 교과서에 쓰이면서 행렬식이라는 용어가 되었다.

만일 $ad - bc = 0$인 경우 식 (1-2-3)은

$$0 \cdot y = aQ - cP$$

이므로 $aQ - cP \neq 0$이면 이 식을 만족하는 y의 값은 없다. 즉, 이 연립방정식은 해를 갖지 않는데 이 경우를 '불능'이라고 한다.

여기서 $aQ - cP = 0$이면 식 (1-2-3)은

$$0 \cdot y = 0$$

이 되어, y에 어떤 수를 넣어도 성립한다. 즉, 해가 무수히 많이 생긴다. 이 경우를 '부정'이라고 한다. 이 경우 $Q = \frac{c}{a}P$로 쓸 수 있는데 이것을 식 (1-2-2)에 넣으면

$$cx + dy = \frac{c}{a}P$$

가 된다. $ad - bc = 0$으로부터 $d = \frac{bc}{a}$이니까 위 식은

$$cx + \frac{bc}{a}y = \frac{c}{a}P$$

로 쓸 수 있다. 양변을 c로 나누고 a를 곱하면

$$ax + by = P$$

가 되어 식 (1-2-1)과 완전히 일치한다.

만일 $ad - bc \neq 0$인 경우 양변을 $ad - bc$로 나누어 y를 구하면

$$y = \frac{aQ - cP}{ad - bc}$$

이고 이 결과를 식 (1-2-1)에 넣으면

$$x = \frac{dP - bQ}{ad - bc}$$

가 된다. 이것을 행렬식으로 다음과 같이 나타낼 수 있다.

$$x = \frac{\begin{vmatrix} P & b \\ Q & d \end{vmatrix}}{\begin{vmatrix} a & b \\ c & d \end{vmatrix}} \qquad y = \frac{\begin{vmatrix} a & P \\ c & Q \end{vmatrix}}{\begin{vmatrix} a & b \\ c & d \end{vmatrix}} \qquad (1\text{-}2\text{-}5)$$

이 식을 크라메르의 공식이라고 부른다.

크라메르는 자신의 방법을 미지수가 3개인 연립방정식에도 적용했다.

$$ax + by + cz = P$$

$$dx + ey + fz = Q$$

$$gx + hy + kz = R$$

크라메르는 이 연립방정식의 계수에 대한 행렬식을

$$\begin{vmatrix} a & b & c \\ d & e & f \\ g & h & k \end{vmatrix} = a \begin{vmatrix} e & f \\ h & k \end{vmatrix} - b \begin{vmatrix} d & f \\ g & k \end{vmatrix} + c \begin{vmatrix} d & e \\ g & h \end{vmatrix}$$

로 정의했다. 그는 이 행렬식이 0이 아닐 때 연립방정식의 해가

$$x = \frac{\begin{vmatrix} P & b & c \\ Q & e & f \\ R & h & k \end{vmatrix}}{\begin{vmatrix} a & b & c \\ d & e & f \\ g & h & k \end{vmatrix}} \quad y = \frac{\begin{vmatrix} a & P & c \\ d & Q & f \\ g & R & k \end{vmatrix}}{\begin{vmatrix} a & b & c \\ d & e & f \\ g & h & k \end{vmatrix}} \quad z = \frac{\begin{vmatrix} a & b & P \\ d & e & Q \\ g & h & R \end{vmatrix}}{\begin{vmatrix} a & b & c \\ d & e & f \\ g & h & k \end{vmatrix}} \quad (1-2-6)$$

로 주어진다는 것을 알아냈다.

정교수 행렬을 이용해 연립방정식을 푸는 새로운 방법을 찾아낸 수학자도 있어.

물리군 그게 누구인가요?

정교수 수학의 왕이라고 불리는 가우스야. 이제 가우스에 대해 자세히 이야기해 보겠네.

가우스(Johann Carl Friedrich Gauss, 1777~1855)

가우스는 1777년 독일에서 가난한 집안의 외동아들로 태어났다. 그의 아버지는 벽돌공과 정원사로 일했는데

성질이 워낙 난폭해서 가족들에게 환영을 받지 못했다. 가우스는 어린 시절 외삼촌 프리드리히에게 수학을 배웠다. 베 짜는 일을 했던 프리드리히는 수학을 좋아해 조카에게 자신이 알고 있는 수학을 모두 가르쳐 주었다.

일곱 살 때 가우스는 성 카타리넨 학교에 입학했다. 그 학교에서는 여러 학년 아이들이 함께 수학을 배웠는데, 교장인 게오르크 뷔트너가 수학을 가르쳤다. 어느 날 뷔트너는 고학년 아이들을 지도하기 위해 저학년 아이들에게 1부터 100까지의 자연수를 모두 더하라는 문제를 냈다.

계산을 마친 아이들은 노트를 교탁 위에 올려놓아야 했는데, 문제를 내준 지 몇 초 만에 가우스가 노트를 가지고 나왔다. 뷔트너는 가우스가 과제를 포기하고 아무렇게나 답을 썼다고 생각하며 화가 난 표정으로 고학년 아이들을 지도했다.

수업이 끝난 뒤 뷔트너는 저학년 아이들의 노트를 들여다보았다. $1 + 2 = 3, 1 + 2 + 3 = 6, \cdots$ 이런 식으로 열심히 덧셈을 한 아이들의 노트 가운데 정답 5050만 적은 가우스의 노트가 있었다.

뷔트너는 가우스에게 어떻게 정답을 알았는지 물었다. 그러자 가우스는 다음과 같이 대답했다.

"1과 100, 2와 99, 3과 98처럼 두 수를 짝지으면 그 합이 항상 101이 됩니다. 이런 짝이 50개가 있으므로 답은 50과 101의 곱인 5050입니다."

가우스의 천재성에 감명받은 뷔트너는 가우스가 더 높은 수준의 수학을 배우기 위해 중학교에 조기 진학하도록 힘써주었다. 14세에 중등교육을 모두 마친 가우스는 브라운슈바이크 공작의 후원으로 과학 아카데미에서 공부한 후 18세에 괴팅겐 대학에 입학했다.

대학 시절 가우스는 수많은 발명을 했다. 19세 때 그는 언어학과 수학 사이에서 전공 결정을 망설였다. 그러다 고대 그리스부터 오랫동안 불가능하다고 여겨진, 정십칠각형을 자와 컴퍼스만으로 작도하는 방법을 알아냈다. 그 후 그는 정257각형, 정65537각형도 똑같은 방법으로 그릴 수 있음을 발견했다.

가우스는 수학 일기를 쓴 것으로 유명하다. 수학 일기는 그가 죽은 후 유족들이 찾아냈는데, 자신이 새롭게 발견한 내용을 한두 줄로 요약한 것이었다. 예를 들어 1796년 3월 30일 일기에는 '원을 17등분할 수 있음'이라는 메모가 남아 있다. 1796년 7월 10일 일기에는 '유레카! 수 = △ + △ + △'라는 암호가 적혀 있는데, 이는 임의의 정수를 세 삼각수의 합으로 나타낼 수 있다는 것을 의미했다. 1796년 10월 21일 일기에는 '나는 거인을 정복했다'고 쓰여 있는데 이것이 무엇을 뜻하는지는 알려지지 않았다.

가우스의 일기장

소행성 케레스의 궤도를 정확하게 계산하는 등 천문학에도 관심이 많았던 가우스는 1807년부터 괴팅겐 대학 교수와 천문대장을 겸임했다. 천문대장이라고 해도 조수 한 명 없이 혼자 천체를 관측하고 계산을 하면서 동시에 강의도 해야 했다. 당시 독일은 프랑스의 나폴레옹이 점령한 상태였기 때문에 점령지의 수학자인 가우스는 아주 적은 급료를 지급받았다. 그는 힘든 생활을 극복하고 전기와 자기 분야에서 수많은 연구를 했다. 전기에 대한 가우스 법칙이 나온 것도 이때의 일이다.

가우스는 말년에 제자인 리만과 함께 새로운 기하학을 만드는 일에 뛰어들었다. 유클리드 기하학은 평면에서만 적용되는데 가우스와 리만은 수박 겉면과 같은 구면에서 적용되는 새로운 기하학을 만든 것이다. 가령 평면에 삼각형을 그리면 세 각의 합이 180도이지만, 구면에 삼각형을 그리면 세 각의 합이 180도보다 커진다는 것이다. 이 새로운 기하학은 가우스가 죽은 뒤 리만이 본격적으로 연구해 리만 기하학이라는 이름으로 불리게 된다.

물리군 연립방정식에 대한 가우스의 방법을 소개해 주세요.

정교수 좋아. 다음 연립방정식을 볼까?

$$\begin{cases} x + 2y = 1 & \cdots ① \\ -x + y = 2 & \cdots ② \end{cases} \quad (1\text{-}2\text{-}7)$$

이 연립방정식을 다음과 같이 행렬로 나타낼 수 있어.

$$\begin{pmatrix} 1 & 2 \\ -1 & 1 \end{pmatrix} \begin{pmatrix} x \\ y \end{pmatrix} = \begin{pmatrix} 1 \\ 2 \end{pmatrix}$$

이때 행렬 $\begin{pmatrix} 1 & 2 \\ -1 & 1 \end{pmatrix}$을 계수행렬이라고 불러. 이번에는 다음 행렬을 만들어 봐.

$$\begin{pmatrix} 1 & 2 & 1 \\ -1 & 1 & 2 \end{pmatrix} \qquad (1\text{-}2\text{-}8)$$

이 행렬은 첨가행렬이라고 하지. 연립방정식 (1-2-7)에서 ①과 ②를 더하면

$$3y = 3 \quad \cdots \text{③}$$

이 돼. 그러니까 ①과 ②를 연립하는 대신 ①과 ③을 연립해 풀 수 있어. 즉, ①과 ②로 이루어진 연립방정식 (1-2-7)은 ①과 ③으로 이루어진 연립방정식으로 바뀌는 거야.

$$\begin{cases} x + 2y = 1 & \cdots \text{①} \\ 3y = 3 & \cdots \text{③} \end{cases} \qquad (1\text{-}2\text{-}9)$$

이 연립방정식에 대한 첨가행렬은

$$\begin{pmatrix} 1 & 2 & 1 \\ 0 & 3 & 3 \end{pmatrix} \qquad (1\text{-}2\text{-}10)$$

이야. 주어진 첨가행렬 (1-2-8)에서 두 행의 합을 제2행으로 대신

할 수 있지. 이렇게 변형하여 y를 구하면 x도 구할 수 있어. 이 경우 $y = \frac{3}{3} = 1$이므로 $x = -1$이 돼. 즉, 첨가행렬 (1-2-8)과 (1-2-10)은 같은 연립방정식을 묘사해. 가우스는 첨가행렬을 조작함으로써 연립방정식을 쉽게 푸는 방법을 알아냈는데, 이것을 가우스 소거법이라고 불러.

행렬의 등장 _ 직사각형 모양으로 수 배열하기

정교수 이제 행렬(Matrix)이 어떻게 탄생했는지 이야기할게. 행렬은 1850년 영국의 수학자 실베스터에 의해 등장했어.

실베스터(James Joseph Sylvester, 1814~1897)

실베스터는 수를 직사각형 모양으로 배열한 행렬을 처음 도입했다.[2]

2] 행렬이라는 이름은 5년 후 실베스터의 친구인 수학자 케일리가 처음 사용했다.

그는 괄호 안에 직사각형 모양으로 수를 배열하여 행렬을 만들었다. 예를 들어 다음 두 행렬을 보자.

$$A = \begin{pmatrix} 2 & 4 \\ 1 & 2 \end{pmatrix}$$

$$B = \begin{pmatrix} 1 \\ 3 \end{pmatrix}$$

행렬 A는 가로로 2줄과 세로로 2줄로 이루어져 있는데 이를 2×2 행렬이라고 부른다. 마찬가지로 행렬 B는 2×1 행렬이다. 이때 가로줄을 행, 세로줄을 열이라고 한다.

일반적으로 행렬 A의 제i행 제j열 원소를 a_{ij}로 나타낸다.

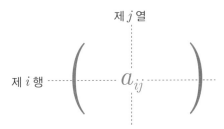

실베스터는 일반적으로 m행과 n열로 이루어진 행렬을 생각했고 이를 $m \times n$ 행렬이라고 불렀다.

또한 두 행렬이 같은 $m \times n$ 행렬이면 더하거나 뺄 수 있고, 그 결과는 $m \times n$ 행렬이 된다고 정의했다. 예를 들어 다음 두 행렬을 보자.

$$A = \begin{pmatrix} a_{11} & a_{12} \\ a_{21} & a_{22} \end{pmatrix}, \quad B = \begin{pmatrix} b_{11} & b_{12} \\ b_{21} & b_{22} \end{pmatrix}$$

두 행렬은 모두 2×2 행렬이다. 이 행렬은 정사각형 모양으로 수를 배열했으므로 정사각행렬이라고도 부르는데, 2×2 행렬을 2차 정사각행렬이라고 한다.

이때 두 행렬의 덧셈과 뺄셈을 다음과 같이 정의한다.

$$A + B = \begin{pmatrix} a_{11} + b_{11} & a_{12} + b_{12} \\ a_{21} + b_{21} & a_{22} + b_{22} \end{pmatrix}$$

$$A - B = \begin{pmatrix} a_{11} - b_{11} & a_{12} - b_{12} \\ a_{21} - b_{21} & a_{22} - b_{22} \end{pmatrix}$$

주어진 행렬 A의 k배를 kA라고 쓰면, kA의 제i행 제j열 원소는 행렬 A의 제i행 제j열 원소의 k배이다.

$$kA = \begin{pmatrix} ka_{11} & ka_{12} \\ ka_{21} & ka_{22} \end{pmatrix}$$

두 행렬 A, B에 대해 두 행렬의 곱셈이 항상 정의되는 것은 아니다. A가 $m \times n$ 행렬이고 B가 $n \times p$ 행렬일 때만 AB가 정의되며, 이때 AB는 $m \times p$ 행렬이 된다. 행렬 A의 제i행 제j열 원소를 a_{ij}라고 하면

$$i = 1, 2, \cdots, m$$

$$j = 1, 2, \cdots, n$$

이고, 행렬 B의 제i행 제j열 원소를 b_{ij}라고 하면

$i = 1, 2, \cdots, n$

$j = 1, 2, \cdots, p$

이다. 이때 AB의 제i행 제j열 원소를 $(AB)_{ij}$라고 하면

$$(AB)_{ij} = \sum_{k=1}^{n} a_{ik} b_{kj}$$

로 정의한다.

다음 두 정사각행렬을 가지고 설명하겠다.

$$A = \begin{pmatrix} a_{11} & a_{12} \\ a_{21} & a_{22} \end{pmatrix}, \quad B = \begin{pmatrix} b_{11} & b_{12} \\ b_{21} & b_{22} \end{pmatrix}$$

두 행렬은 모두 2×2 행렬이니까 AB가 정의된다. 두 행렬의 곱을 구하면

$$AB = \begin{pmatrix} a_{11}b_{11} + a_{12}b_{21} & a_{11}b_{12} + a_{12}b_{22} \\ a_{21}b_{11} + a_{22}b_{21} & a_{21}b_{12} + a_{22}b_{22} \end{pmatrix}$$

이다. 즉,

$$(AB)_{11} = a_{11}b_{11} + a_{12}b_{21} = \sum_{l=1}^{2} a_{1l} b_{l1}$$

$$(AB)_{12} = a_{11}b_{12} + a_{12}b_{22} = \sum_{l=1}^{2} a_{1l}b_{l2}$$

$$(AB)_{21} = a_{21}b_{11} + a_{22}b_{21} = \sum_{l=1}^{2} a_{2l}b_{l1}$$

$$(AB)_{22} = a_{21}b_{12} + a_{22}b_{22} = \sum_{l=1}^{2} a_{2l}b_{l2}$$

가 된다.

다음 두 행렬을 예로 들어 보자.

$$A = \begin{pmatrix} 1 & 4 \\ 2 & 3 \end{pmatrix}, \quad B = \begin{pmatrix} -1 & 2 \\ 0 & 4 \end{pmatrix}$$

이때

$$AB = \begin{pmatrix} -1 & 18 \\ -2 & 16 \end{pmatrix}$$

이고,

$$BA = \begin{pmatrix} 3 & 2 \\ 8 & 12 \end{pmatrix}$$

이다. 두 결과를 비교하면 $AB \neq BA$임을 알 수 있다. 즉, 일반적으로 행렬의 곱은 교환법칙을 만족하지 않는다.

대각행렬 중에서 대각성분이 모두 같은 수일 때 이 행렬을 스칼라 행렬이라고 부른다. 스칼라 행렬 중에서 대각성분이 모두 1이면 단위행렬이라 하고 I로 쓴다.

행렬 A, B, C에 대해 다음이 성립한다.

(1) $(kA)B = k(AB)$　(k는 실수)
(2) $A(BC) = (AB)C$　(결합)
(3) $A(B+C) = AB + AC$　(분배)
(4) $C(A+B) = CA + CB$　(분배)

한편 정사각행렬 A에 대해 $\underbrace{AAA\cdots A}_{n}$를 A^n으로 나타내고 행렬의 거듭제곱이라고 부른다.

예를 들어 다음 행렬을 보자.

$$A = \begin{pmatrix} 1 & a \\ 0 & 1 \end{pmatrix}$$

이 행렬의 거듭제곱을 구하면

$$A^n = \begin{pmatrix} 1 & na \\ 0 & 1 \end{pmatrix}$$

이 된다.

이제 행렬의 전치에 대해 알아보자. 전치는 영어로 transpose이다. '전'은 바꾼다는 뜻이고 '치'는 위치를 뜻한다. 즉, 전치란 위치를 바꾸는 것을 말한다.

물리군   어떤 위치를 바꾸는 거죠?

정교수   행과 열을 바꾸는 것을 전치라고 해. 어떤 행렬 A를 전치시킨 행렬을 그 행렬의 전치행렬이라 하고 A^T라고 써. 예를 들어 다음 행렬을 봐.

$$B = \begin{pmatrix} 1 & 2 \\ 3 & 4 \end{pmatrix}$$

제1행은 1 2이고 제2행은 3 4이지? 1 2가 제1열이고 3 4가 제2열인 행렬이 이 행렬의 전치행렬이야. 즉, 다음과 같아.

$$B^T = \begin{pmatrix} 1 & 3 \\ 2 & 4 \end{pmatrix}$$

이번엔 다음 행렬의 전치행렬을 구해 볼까?

$$B = \begin{pmatrix} 1 \\ 2 \end{pmatrix}$$

물리군   그건 간단해요.

$$B^T = (1 \quad 2)$$

정교수   여기서

$$BB^T = \begin{pmatrix} 1 & 2 \\ 2 & 4 \end{pmatrix}$$

이고,

$$B^T B = 5$$

가 돼.

물리군 두 행렬을 곱했는데 수가 나오네요.

정교수 B^T는 1×2 행렬이고 B는 2×1 행렬이니까 $B^T B$는 1×1 행렬, 즉 수가 되지.

전치에는 다음과 같은 성질이 있어.

(1) $(A^T)^T = A$

(2) $(A+B)^T = A^T + B^T$

(3) $(kA)^T = kA^T$

(4) $(AB)^T = B^T A^T$

물리군 (4)에서는 순서가 바뀌는군요.

정교수 $N \times N$ 정사각행렬로 증명해 볼게.

$(AB)^T$의 제i행 제j열 원소를 $[(AB)^T]_{ij}$라고 하면

$$[(AB)^T]_{ij} = (AB)_{ji}$$

이다. 행렬곱의 정의에 따라

$$(AB)_{ji} = \sum_{k=1}^{N} a_{jk} b_{ki}$$

이고

$$a_{jk} = (A^T)_{kj}, \quad b_{ki} = (B^T)_{ik}$$

이니까

$$(AB)_{ji} = \sum_{k=1}^{N} a_{jk} b_{ki} = \sum_{k=1}^{N} (B^T)_{ik} (A^T)_{kj} = [B^T A^T]_{ij}$$

가 된다.

물리군   이제 이해했어요.

정교수   그럼 정사각행렬에 대한 역행렬을 알아볼까? 행렬 A가 정사각행렬이라고 할 때 이 행렬의 역행렬은 A^{-1}로 쓰는데

$$AA^{-1} = A^{-1}A = I$$

를 만족해. 두 행렬의 곱의 역행렬은 다음과 같은 성질이 있지.

$$(AB)^{-1} = B^{-1} A^{-1}$$

벡터의 탄생 _ 오래 누워 있는 습관 덕에

정교수 지금부터는 벡터의 탄생 과정을 알아보기로 하세. 벡터는 데카르트가 좌표를 발견하면서부터 탄생하기 시작했다고 볼 수 있어. 그럼 데카르트에 대해 먼저 소개할게.

데카르트(René Descartes, 1596~1650)

데카르트는 프랑스 라에에서 태어났다. 그의 어머니는 데카르트를 낳고 1년 1개월 후에 죽었다. 어릴 때부터 약골이었던 데카르트는 라 플레슈의 기독교 학교에 입학했지만 몸이 약해 수업을 제대로 듣지 못하고 자주 늦잠을 자곤 했다. 교장인 샤를레 신부는 데카르트를 마음에 들어 했는데, 그에게 아침 수업에 들어오지 않고 침대에 누워 있을 수 있게 해주었다. 이렇게 침대에 오래 누워 있는 습관 덕에 데카르트는 날벌레가 천장에 붙어 있는 것을 보고 그 위치를 계산하려다가 좌표를 만들었다는 일화도 있다.

1614년 데카르트는 푸아티에 대학에 입학해 법학과 의학을 공부했다. 대학 졸업 후 그는 세상을 배우기 위해서 네덜란드의 마우리츠 공 휘하에서 군인 생활을 했다. 1617년 어느 날 그는 거리에 걸려 있는 네덜란드어로 쓰인 글을 보고 지나가던 행인에게 그 내용을 프랑스어로 번역해 달라고 부탁했다. 그 행인은 홀란트의 대학 학장이자 수학자였던 이사크 베이크만(Isaac Beeckman)이었다.

베이크만은 데카르트에게 "내가 내는 수학 문제를 풀면 번역해 주겠다"고 제안했다. 그 문제는 당시 아무도 풀지 못하는 문제였다. 데카르트는 몇 시간 만에 문제를 풀어 베이크만을 놀라게 했다. 이 일을 계기로 베이크만은 데카르트에게 학문을 하라고 권유했다.

1619년 어느 추운 겨울날 데카르트는 군대 막사에서 재미있는 수학 문제를 떠올렸다. 그것은 다면체에서 점의 개수를 v, 변의 개수를 e, 면의 개수를 f라고 하면

$$v - e + f = 2$$

가 된다는 것이다. 이 공식은 데카르트가 먼저 발견했지만 훗날 오일러의 공식으로 알려졌다. 하지만 수학자들은 이를 데카르트-오일러 공식이라고 부른다.

1621년에 군 생활을 그만둔 데카르트는 독일, 덴마크, 네덜란드, 스위스, 이탈리아를 여행하면서 순수수학과 물리학, 철학을 연구했다. 그가 책을 쓴 것은 네덜란드에서 체류하는 20년 동안이었다.

데카르트는 1637년에 철학책인 《방법 서설(Discours de la méthode)》

을 출간했다. 그는 이 책에서 '나는 생각한다. 고로 존재한다.'라는 유명한 말을 남겼다.

《방법 서설》

같은 해 데카르트는 그의 위대한 수학책인 《기하학(La Géométrie)》을 출간했다.

《기하학》의 첫 페이지

데카르트의 수학 연구 내용은 이 책에 거의 대부분 들어 있다. 그는 방정식에서 미지수를 x, y, z 등으로 나타내기 시작했다. 또한 x^2, x^3과 같은 거듭제곱 표현을 처음 사용했다.

《기하학》은 총 세 권으로 이루어져 있다. 1권에서 데카르트는 평면 위의 점의 좌표를 표시하는 방법을 최초로 도입했다.

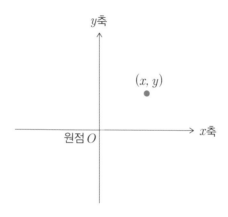

이렇게 나타내는 좌표를 데카르트 좌표라고 부른다. 데카르트는 좌표를 이용해 원, 포물선, 타원, 쌍곡선을 정의하고, 이들 도형의 많은 성질을 좌표를 가지고 증명할 수 있었다.

정교수 이번에는 벡터를 연구한 독일 수학자 그라스만의 이야기를 해볼게.

그라스만(Hermann Günther Grassmann, 1809~1877)

그라스만의 어머니는 장관의 딸이었고, 그라스만의 아버지는 장관을 지내다가 슈테틴 김나지움의 교사가 되어 수학과 물리학을 가르쳤다. 1827년 그라스만은 베를린 대학에서 신학과 고전 학문을 공부한 후 교회에서 일자리를 얻었다.

교사가 되려고 준비하던 그라스만은 밀물과 썰물에 관한 논문을 썼고 이를 위해 라플라스의 천체역학을 공부했다. 1844년 그는 〈선형 확장 이론〉이라는 논문을 발표했다. 이 내용은 1862년에 《확장 이론》이라는 제목의 책으로 출판되었다. 이 책에서 그라스만은 벡터 공간과 선형의 개념을 도입했다. 또한 3차원보다 더 높은 차원에서의 벡터도 다루었다.

물리군 벡터는 어떻게 정의하나요?

정교수 그림을 가지고 차근차근 설명해 줄게.

2차원 평면 위의 한 점 $P(3, 2)$를 생각하자. 이때 원점 O를 꼬리로, 점 P를 머리로 하는 벡터 $\overrightarrow{OP}$는 다음 그림과 같다.

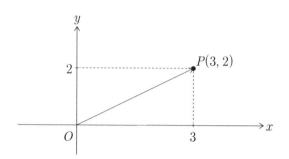

이 벡터를 점 P의 위치벡터라고 부른다. 점 P의 위치벡터를 다음과 같이 두 벡터의 합으로 나타내자.

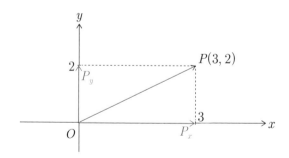

$$\overrightarrow{OP} = \overrightarrow{OP}_x + \overrightarrow{OP}_y \qquad (1\text{-}4\text{-}1)$$

$\overrightarrow{OP_x}$는 x축에 나란하고 $\overrightarrow{OP_y}$는 y축에 나란하다. $\overrightarrow{OP_x}$의 크기는 3이고 $\overrightarrow{OP_y}$의 크기는 2이다. 이제 $\overrightarrow{OP_x}$와 나란하면서 크기가 1인 벡터를 $\hat{i}$라고 하면

$$\hat{i} = \frac{\overrightarrow{OP_x}}{3} \qquad (1\text{-}4\text{-}2)$$

이고, 마찬가지로 $\overrightarrow{OP_y}$와 나란하면서 크기가 1인 벡터를 $\hat{j}$라고 하면

$$\hat{j} = \frac{\overrightarrow{OP_y}}{2} \qquad (1\text{-}4\text{-}3)$$

이다. 식 (1-4-2)와 (1-4-3)에서

$$\overrightarrow{OP_x} = 3\hat{i}$$

$$\overrightarrow{OP_y} = 2\hat{j}$$

이므로

$$\overrightarrow{OP} = \overrightarrow{OP_x} + \overrightarrow{OP_y} = 3\hat{i} + 2\hat{j}$$

가 된다. 이때 3을 $\overrightarrow{OP}$의 x성분, 2를 $\overrightarrow{OP}$의 y성분이라고 부른다. 한편 여기서 다음과 같은 사실을 알 수 있다.

$-\hat{i}$는 크기가 1이고 $\hat{i}$와 방향이 반대인 벡터이다.

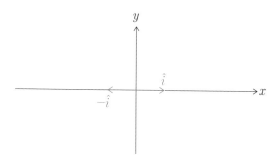

$-\hat{j}$는 크기가 1이고 $\hat{j}$와 방향이 반대인 벡터이다.

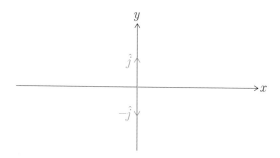

이제 2차원 평면 위의 한 점 $P(x, y)$를 생각하자. 이때 원점 O를 꼬리로, 점 P를 머리로 하는 벡터 $\overrightarrow{OP}$는 다음 그림과 같다.

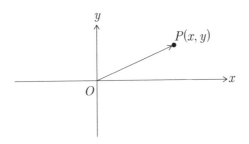

이 벡터를 점 P의 위치벡터라고 부른다. 이 벡터는 너무 자주 나타나니까 다음과 같이 쓴다.

$$\vec{r} = \overrightarrow{OP}$$

앞으로 $\vec{r}$의 크기를 r이라고 쓰는데 이것은 원점으로부터 점 P까지의 거리이다. 그러니까

$$r = \sqrt{x^2 + y^2}$$

이 된다. 이 벡터는 다음과 같이 쓸 수 있다.

$$\vec{r} = \overrightarrow{OP}$$
$$= \overrightarrow{OP_x} + \overrightarrow{OP_y}$$
$$= x\hat{i} + y\hat{j}$$

여기서 x를 $\overrightarrow{OP}$의 x성분, y를 $\overrightarrow{OP}$의 y성분이라고 부른다.

꼬리가 원점이 아닌 임의의 벡터 $\vec{A}$는 꼬리가 원점이 되도록 평행이동 할 수 있다.

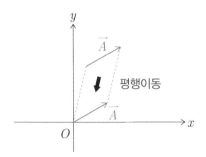

따라서

$$\vec{A} = A_x \hat{i} + A_y \hat{j}$$

로 나타낼 수 있다. 이때 A_x를 $\vec{A}$의 x성분, A_y를 $\vec{A}$의 y성분이라고 부른다.

$\vec{A}$의 크기($|\vec{A}|$)는 A_x, A_y를 두 변으로 하는 직각삼각형의 빗변의 길이이므로

$$|\vec{A}| = \sqrt{A_x^2 + A_y^2}$$

이 된다.

이제 다음 두 벡터를 보자.

$$\vec{A} = A_x \hat{i} + A_y \hat{j}$$

$$\vec{B} = B_x \hat{i} + B_y \hat{j}$$

이때 두 벡터의 덧셈과 뺄셈은 다음과 같다.

$$\vec{A} + \vec{B} = (A_x + B_x)\hat{i} + (A_y + B_y)\hat{j}$$

$$\vec{A} - \vec{B} = (A_x - B_x)\hat{i} + (A_y - B_y)\hat{j}$$

벡터의 k배는 다음과 같다.

$$k\vec{A} = kA_x \hat{i} + kA_y \hat{j}$$

이제 벡터의 내적을 알아보자.

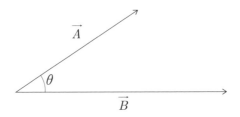

여기서 θ는 두 벡터의 사잇각이다. 두 벡터 $\vec{A}$와 $\vec{B}$의 내적은 다음과 같이 정의한다.

$$\vec{A} \cdot \vec{B} = |\vec{A}||\vec{B}|\cos\theta$$

내적은 교환법칙을 만족한다.

$$\vec{A} \cdot \vec{B} = \vec{B} \cdot \vec{A}$$

내적의 정의로부터 θ가 작아질수록 $\cos\theta$가 커지므로 $\theta = 0$일 때 두 벡터의 내적이 가장 크다. $\theta = 0$이라는 것은 두 벡터가 평행하다는 뜻이다. 극단적으로 같은 두 벡터 $\vec{A}$의 경우 $\theta = 0$이니까 $\cos\theta = 1$이므로

$$\vec{A} \cdot \vec{A} = |\vec{A}|^2$$

이 된다. 단위벡터는 크기가 1이니까 같은 단위벡터끼리의 내적은 1이다.

$$\hat{i} \cdot \hat{i} = 1 \qquad \hat{j} \cdot \hat{j} = 1$$

$\theta = \dfrac{\pi}{2}$이면 $\cos\theta = 0$이므로 두 벡터가 이루는 각이 $\dfrac{\pi}{2}$이면 내적이 0이다. $\dfrac{\pi}{2}$를 각도 단위로 쓰면 90°이니까 수직으로 만나는 두 벡터 $\vec{A}$와 $\vec{B}$의 내적은 0이다.

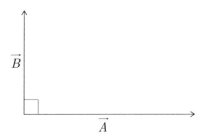

따라서 수직으로 만나는 단위벡터끼리의 내적은 0임을 알 수 있다.

$\hat{i} \cdot \hat{j} = 0$

다음 두 벡터를 보자.

$\vec{A} = A_x \hat{i} + A_y \hat{j}$

$\vec{B} = B_x \hat{i} + B_y \hat{j}$

두 벡터의 내적은 다음과 같다.

$\vec{A} \cdot \vec{B} = A_x B_x + A_y B_y$

물리군 어떻게 이런 결과가 나왔죠?
정교수 직접 계산해 볼까?

$$\vec{A} \cdot \vec{B} = (A_x \hat{i} + A_y \hat{j}) \cdot (B_x \hat{i} + B_y \hat{j})$$

$$= A_x B_x \hat{i} \cdot \hat{i} + A_x B_y \hat{i} \cdot \hat{j} + A_y B_x \hat{j} \cdot \hat{i} + A_y B_y \hat{j} \cdot \hat{j}$$

$$= A_x B_x + A_y B_y$$

이것을 이용하면 두 벡터가 같을 때 벡터의 크기를 알 수 있다. $\vec{B}$자리에 $\vec{A}$를 넣으면

$$\vec{A} \cdot \vec{A} = A_x^2 + A_y^2$$

이니까 이것과 $\vec{A} \cdot \vec{A} = |\vec{A}|^2$을 이용하면

$$|\vec{A}|^2 = A_x^2 + A_y^2$$

이다. 이 식의 양변에 제곱근을 취하면 벡터의 크기는

$$|\vec{A}| = \sqrt{A_x^2 + A_y^2}$$

이 된다.

세 개의 벡터 $\vec{A}, \vec{B}, \vec{C}$에 대해서는 다음이 성립한다.

$$\vec{A} \cdot (\vec{B} + \vec{C}) = \vec{A} \cdot \vec{B} + \vec{A} \cdot \vec{C}$$

그라스만의 벡터 공간 _체를 이루는 조건

정교수 이제 벡터 공간에 대해서 살펴볼게. 벡터 공간을 이해하려면 체(Field)에 관해 조금 알 필요가 있어.

체 F는 다음 조건을 만족하는 집합이다.
(1) 체 F의 임의의 원소 a에 대해 $a + 0 = a$를 만족하는 0이 체 F 속에 있다.
(2) 체 F의 임의의 원소 a에 대해 $a \times 1 = a$를 만족하는 1이 체 F 속에 있다.
(3) 체 F의 각 원소 a에 대해 $a + (-a) = 0$인 $-a$가 체 F 속에 있다.
(4) 체 F의 각 원소 a에 대해 $a \times a^{-1} = 1$인 a^{-1}이 체 F 속에 있다.
이때 F가 수 집합이면 이 체를 수체라고 부른다.

물리군 자연수의 집합은 체인가요?

정교수 자연수의 집합은 0이 없으므로 체를 이루지 않아.

물리군 그럼 정수의 집합은요?

정교수 정수의 집합은 $2 \times a^{-1} = 1$인 a^{-1}이 존재하지 않으므로 역시 체를 이루지 않지. 0으로 나누는 것을 금지한다면 유리수 집합이나 실수 집합, 복소수 집합은 체를 이룬다네.

물리군 벡터 공간은 어떤가요?

정교수 벡터 공간은 다음 성질을 만족해야 해.

다음과 같은 성질을 만족하는 집합 V를 체 F 위에서의 벡터 공간이라고 정의한다. $\vec{a}, \vec{b}, \vec{c}, \vec{u}, \vec{v}$ 등은 집합 V의 원소이다.

(1) $\vec{a} + (\vec{b} + \vec{c}) = (\vec{a} + \vec{b}) + \vec{c}$

(2) $\vec{a} + \vec{b} = \vec{b} + \vec{a}$

(3) 집합 V의 모든 원소 $\vec{v}$에 대해 $\vec{v} + \vec{0} = \vec{v}$인 $\vec{0}$이 집합 V에 속한다.

(4) 집합 V의 모든 원소 $\vec{v}$에 대해 $\vec{v} + (-\vec{v}) = \vec{0}$인 $-\vec{v}$가 집합 V에 속한다.

(5) 체 F의 원소 k, m에 대해 다음이 성립한다.
$k(m\vec{v}) = (km)\vec{v}$

(6) 체 F의 원소 k에 대해 다음이 성립한다.
$k(\vec{u} + \vec{v}) = k\vec{u} + k\vec{v}$

(7) 체 F의 원소 k, m에 대해 다음이 성립한다.
$(k + m)\vec{v} = k\vec{v} + m\vec{v}$

이러한 벡터 공간의 원소를 벡터라고 부른다.

물리군 우리가 흔히 알고 있는 벡터도 벡터 공간을 이루나요?

정교수 물론이야. 이 벡터 공간을 V라고 할게. 그리고 앞으로

$\hat{i} = \hat{e}_1$

$\hat{j} = \hat{e}_2$

라고 쓸 거야.

이제 다음 식을 볼까?

$$a_1 \hat{e}_1 + a_2 \hat{e}_2 = 0 \qquad (1\text{-}5\text{-}1)$$

이 등식이 성립하기 위한 필요충분조건이 $a_1 = a_2 = 0$이면 $\hat{e}_1$과 $\hat{e}_2$는 일차독립이라고 말해. 그럼 이들이 일차독립이라는 걸 보여줄게. 우선 $a_1 = a_2 = 0$이면 $a_1 \hat{e}_1 + a_2 \hat{e}_2 = 0$이 성립하는 것은 자명해. 이제

$$a_1 \hat{e}_1 + a_2 \hat{e}_2 = 0$$

인 경우를 살펴보세. 이 식의 양변에 $\hat{e}_1$을 내적하면

$$a_1 \hat{e}_1 \cdot \hat{e}_1 + a_2 \hat{e}_1 \cdot \hat{e}_2 = 0$$

이 되어, $a_1 = 0$이야. 같은 방법으로 식 (1-5-1)의 양변에 $\hat{e}_2$를 내적하면 $a_2 = 0$이 되지. 그러니까 2차원에서의 단위벡터 $\hat{e}_1$과 $\hat{e}_2$는 일차독립이라네.

그러면 V의 임의의 원소 $\vec{u}$는

$$\vec{u} = u_1 \hat{e}_1 + u_2 \hat{e}_2 \qquad (1\text{-}5\text{-}2)$$

로 나타낼 수 있어. 이렇게 벡터 공간 V의 모든 원소(벡터)를 표현할 수 있는 $\hat{e}_1$과 $\hat{e}_2$를 기저(base)라고 불러.

물리군 우리가 생각하는 벡터 공간이 2차원이니까 기저가 2개 필요하군요.

정교수 맞아. 기저의 수는 차원만큼 필요해. 3차원 벡터 공간이면 3개의 기저가, 4차원 공간이면 4개의 기저가 필요하지. 그런데 기저 $\hat{e}_1$과 $\hat{e}_2$는 크기가 1이고 서로 수직이야. 이러한 기저를 정규직교기저라고 불러. 정규는 크기가 1, 직교는 수직인 것을 뜻하지.

식 (1-5-2)처럼 벡터 공간의 임의의 원소를 나타내는 것을 일차결합 또는 선형결합이라고 해.

두 번째 만남

디랙의 양자역학

디랙 브라켓의 탄생 _ 벡터를 표현하는 또 다른 기호

정교수   두 번째 만남에서는 디랙이 새로운 표현으로 쓴 양자역학을 알아볼 거야. 반입자 예언으로 노벨 물리학상을 받은 디랙은 이전에 없던 수학적인 방법을 만들었지. 먼저 그가 만든 벡터 표현법을 살펴볼까?

2차원에서의 벡터

$$\vec{A} = 3\hat{e}_1 + 4\hat{e}_2$$

를 생각해 봐. 디랙은 이 벡터를 표현하는 새로운 기호를 도입했어.

물리군   어떤 기호인가요?

정교수   우선 단위벡터를 다음과 같이 행렬로 나타내는 거야.

$$\hat{e}_1 = \begin{pmatrix} 1 \\ 0 \end{pmatrix}$$

$$\hat{e}_2 = \begin{pmatrix} 0 \\ 1 \end{pmatrix}$$

그러면 벡터 $\vec{A}$는

$$\vec{A} = \begin{pmatrix} 3 \\ 4 \end{pmatrix}$$

로 표현할 수 있지. 여기서 디랙은

$$\hat{e}_1 = \begin{pmatrix} 1 \\ 0 \end{pmatrix} = |e_1\rangle$$

$$\hat{e}_2 = \begin{pmatrix} 0 \\ 1 \end{pmatrix} = |e_2\rangle$$

$$\vec{A} = \begin{pmatrix} 3 \\ 4 \end{pmatrix} = |A\rangle$$

로 쓰고 | 〉를 켓벡터라고 불렀어. 즉

$$|A\rangle = 3|e_1\rangle + 4|e_2\rangle$$

가 되지.

물리군 두 개의 켓벡터 $|e_1\rangle$과 $|e_2\rangle$가 기저가 되는군요.

정교수 맞아. 이번에는

$$\vec{B} = 4\hat{e}_1 + 3\hat{e}_2$$

를 봐. 이 벡터는 다음과 같이 켓벡터로 나타낼 수 있어.

$$\vec{B} = |B\rangle = \begin{pmatrix} 4 \\ 3 \end{pmatrix}$$

즉,

$$|B\rangle = 4|e_1\rangle + 3|e_2\rangle$$

가 되지. 그러므로 두 켓벡터의 덧셈을 정의할 수 있다네.

$$|A\rangle + |B\rangle = 7|e_1\rangle + 7|e_2\rangle$$

물리군 두 벡터의 내적은 어떻게 정의하나요?

정교수 내적의 정의로부터

$$\vec{A} \cdot \vec{B} = 24$$

가 돼. 이것을 나타내기 위해 디랙은 켓벡터의 전치행렬이 필요했어. 그래서 다음과 같이 표현했지.

$$\langle\ | = |\ \rangle^T$$

즉,

$$\langle e_1 | = \begin{pmatrix} 1 \\ 0 \end{pmatrix}^T = (1\ \ 0)$$

$$\langle e_2 | = \begin{pmatrix} 0 \\ 1 \end{pmatrix}^T = (0\ \ 1)$$

$$\langle A | = \begin{pmatrix} 3 \\ 4 \end{pmatrix}^T = (3\ \ 4)$$

가 되지. 이것을 정리하면 다음과 같아.

$$\langle A | = 3\langle e_1| + 4\langle e_2|$$

디랙은 두 벡터 $\vec{A}$와 $\vec{B}$의 내적을

$$\vec{A} \cdot \vec{B} = \langle A | B \rangle$$

로 썼는데, 이때

$$\langle A | B \rangle = (3 \ 4) \binom{4}{3} = 24$$

가 되지. 디랙은 켓벡터의 전치인 〈 |를 브라벡터라고 불렀다네. 그러니까 같은 두 벡터의 내적은

$$\vec{A} \cdot \vec{A} = \langle A | A \rangle$$

이므로 벡터 $\vec{A}$의 크기는

$$|\vec{A}| = \sqrt{\langle A | A \rangle}$$

라네.

물리군 켓과 브라라는 이름은 어디에서 나온 거죠?

정교수 〈 〉는 영어로 bracket이야. 여기서 c 왼쪽에는 브라(bra)가 남고 오른쪽에는 켓(ket)만 남기 때문이지.

물리군 재미있는 이름이네요. 브라벡터는 반드시 켓벡터의 왼쪽에 써야 하나요?

정교수 꼭 그렇지는 않아. 브라벡터가 켓벡터의 왼쪽에서 곱해지면 내적으로 수가 되지만 브라벡터가 켓벡터의 오른쪽에서 곱해지면 행렬이 돼. 예를 들어

$$|B\rangle\langle A| = \begin{pmatrix} 4 \\ 3 \end{pmatrix}(3\ 4) = \begin{pmatrix} 12 & 16 \\ 9 & 12 \end{pmatrix}$$

이지.

물리군 행렬의 교환법칙이 성립하지 않아 달라지는군요.

정교수 그렇지. 이번에는 좌표평면 위의 점을 켓벡터로 나타내는 방법을 알려줄게.

2차원 좌표평면 위의 점 $P(x, y)$는 원점 $O(0, 0)$에서 점 P로 향하는 벡터

$$\overrightarrow{OP} = x\hat{e}_1 + y\hat{e}_2$$

로 쓸 수 있다. 그러니까 이 벡터가 되는 켓벡터를 $|X\rangle$로 나타내면

$$|X\rangle = x|e_1\rangle + y|e_2\rangle = \begin{pmatrix} x \\ y \end{pmatrix}$$

이다. 이 벡터에 대응하는 브라벡터는

$$\langle X| = x\langle e_1| + y\langle e_2| = (x\ y)$$

가 된다. 그러므로 원점에서 점 P까지의 거리의 제곱은

$$\langle X|X\rangle = x^2 + y^2$$

으로 나타낼 수 있다.

이번에는 또 다른 켓벡터

$$|X'\rangle = x'|e_1\rangle + y'|e_2\rangle = \begin{pmatrix} x' \\ y' \end{pmatrix}$$

을 보자. 이때

$$\langle X'|X\rangle = x'x + y'y$$

이고,

$$\langle X|X'\rangle = x'x + y'y$$

가 되어,

$$\langle X'|X\rangle = \langle X|X'\rangle$$

이 성립한다.

물리군   그러니까 $|e_1\rangle$과 $|e_2\rangle$가 2차원 벡터 공간의 기저가 되는군요.
정교수   맞아. 여기서

$$\langle e_1|e_2\rangle = 0$$

이니까 $|e_1\rangle$과 $|e_2\rangle$는 서로 수직이야. 이렇게 서로 수직인 기저들을 직교기저 또는 직교 켓벡터라고 부르지. 그리고

$$\langle e_1 | e_1 \rangle = \langle e_2 | e_2 \rangle = 1$$

이니까 $|e_1\rangle$과 $|e_2\rangle$는 크기가 1인 켓벡터야. 이렇게 크기가 1인 켓벡터를 정규화한 켓벡터라고 불러. 그러니까 $|e_1\rangle$과 $|e_2\rangle$는 정규직교화한 켓벡터야. 그리고 $|e_1\rangle$과 $|e_2\rangle$는 다음 성질을 만족해.

$$|e_1\rangle\langle e_1| + |e_2\rangle\langle e_2| = I = \begin{pmatrix} 1 & 0 \\ 0 & 1 \end{pmatrix}$$

물리군 브라켓은 참 흥미로운 기호군요.

정교수 이번에는 조금 더 복잡한 벡터 공간을 생각해 보려고 해. 일반적으로 복소수 $z = x + iy$는 2차원 좌표평면 위의 점 (x, y)에 대응하지. 이때 z의 켤레복소수는 $z^* = x - iy$이고, 원점에서 점 (x, y)까지의 거리의 제곱은

$$zz^* = |z|^2 = x^2 + y^2$$

이 되지. 여기서 $|z|$를 복소수 z의 크기라고 불러.

이제 2차원 복소 공간을 생각할게. 2차원 복소 공간에서는 한 점이 두 개의 복소수로 묘사돼. 예를 들어 임의의 점

$$P(z_1, z_2)$$

를 생각해 볼까? 이때 z_1과 z_2는 복소수야. 2차원 복소 공간의 한 점을 켓벡터로 나타내면 다음과 같아.

$$|Z\rangle = \begin{pmatrix} z_1 \\ z_2 \end{pmatrix}$$

이 켓벡터의 브라벡터는 $(z_1\ z_2)$가 아니야.

물리군   예상과 다르네요?

정교수   2차원 복소 공간에서 원점과 $P(z_1, z_2)$ 사이의 거리는

$$z_1^2 + z_2^2$$

이 아니거든.

물리군   그건 왜죠?

정교수   거리는 복소수가 될 수 없기 때문이야. 거리는 반드시 실수이어야 해. 그러니까 2차원 복소 공간에서 원점과 $P(z_1, z_2)$ 사이의 거리는

$$z_1^* z_1 + z_2^* z_2 = |z_1|^2 + |z_2|^2$$

으로 정의돼. 그러면 두 점 사이의 거리는 실수이면서 동시에 음수가 아니지.

물리군   거리는 음수이면 안 되죠!

정교수   그래서 2차원 복소 공간에서는 브라벡터를 조금 다르게 정의해.

물리군   어떻게 정의하나요?

정교수   켓벡터의 모든 성분을 켤레로 바꾸고 전치한 것을 브라벡터로 정의하지. 이렇게 모든 성분을 켤레로 바꾸고 전치하는 것을 수반

(adjoint)이라고 해. 즉, 성분이 복소수인 경우에는

$$\langle \ | = (| \ \rangle^*)^T$$

가 되는데 이것을

$$\langle \ | = (| \ \rangle)^\dagger$$

로 쓰고 †를 대거(dagger)라고 읽어. 그러니까

$$\langle Z | = (z_1^* \ z_2^*)$$

이므로

$$\langle Z | Z \rangle = z_1^* z_1 + z_2^* z_2 = |z_1|^2 + |z_2|^2$$

이지.

2차원 복소 공간에서 또 다른 켓벡터

$$| W \rangle = \begin{pmatrix} w_1 \\ w_2 \end{pmatrix}$$

를 생각해 봐. 이때

$$\langle W | Z \rangle = w_1^* z_1 + w_2^* z_2$$

이고,

$$\langle Z | W \rangle = z_1^* w_1 + z_2^* w_2$$

가 돼.

물리군 $\langle W|Z \rangle$와 $\langle Z|W \rangle$가 다르네요.

정교수 켓벡터나 브라벡터가 복소수로 표현되었기 때문이야. 이 경우에는 다음과 같은 성질이 있지.

$$\langle W | Z \rangle = \langle Z | W \rangle^*$$

코시 적분 _ 새로운 적분 공식

정교수 이번에는 새로운 적분 공식을 소개하려고 해.

물리군 어떤 공식인데요?

정교수 다음과 같아.

$$\int_0^\infty \frac{\sin x}{x} dx = \frac{\pi}{2} \tag{2-2-1}$$

물리군 처음 보는 적분 형태예요.

정교수 이 적분을 최초로 알아낸 사람은 프랑스의 수학자 코시야.

코시(Augustin-Louis Cauchy, 1789~1857)

코시는 1789년 파리에서 태어났다. 그의 아버지는 파리 경찰 고위 관리였지만, 코시가 태어나기 한 달 전에 발생한 프랑스 혁명(1789년 7월 14일)으로 이 직위를 잃었다. 1794년 코시의 가족은 폭동을 피해, 아르쾨유에 있는 시골집으로 피신했다. 로베스피에르가 처형된 후 가족은 안전하게 파리로 돌아갔고, 코시의 아버지는 새 정부에서 상원의 사무국장이 되었다.

1802년 가을, 코시는 판테온의 에콜 상트랄에 입학해 라틴어와 인문학을 공부했다. 1805년 그는 토목기사가 되기 위해 에콜 폴리테크니크에 입학했다. 1810년에 학교를 졸업한 후에는 나폴레옹이 해군기지를 건설하려고 했던 셰르부르에서 하급 엔지니어로 일하면서 틈틈이 수학을 공부했다.

1812년 9월, 코시는 과로로 병이 나서 파리로 돌아왔다. 파리에서 그는 대칭함수, 대칭군, 고차 대수방정식 이론 등을 연구했다. 1814

년에는 복소함수론의 아이디어를 냈다. 1815년 코시는 에콜 폴리테크니크의 수학 교수가 되었고, 이때부터 본격적으로 복소함수 연구를 시작했다.

물리군   코시가 이 적분을 어떻게 계산했나요?
정교수   그것을 이해하려면 수학과 3학년 때 배우는 복소변수함수론이라는 수학이 필요해. 그 내용을 설명하는 대신 조금 더 쉬운 방법으로 식 (2-2-1)을 증명해 볼게. 이 증명은 이 책의 주인공인 파인먼의 방법이야.

우선 다음 적분을 알아야 한다.

$$A = \int \frac{1}{1+x^2} dx$$

이 식에서

$$x = \tan\theta$$

로 치환하면

$$dx = \sec^2\theta d\theta$$

이고

$$1 + x^2 = 1 + \tan^2\theta = \sec^2\theta$$

이니까

$$A = \int d\theta = \theta + C$$

가 된다. $x = \tan\theta$일 때

$$\theta = \tan^{-1} x$$

라고 쓴다. 여기서 $\tan^{-1} x$는 $\tan x$의 역함수이다.

$$\tan 0 = 0$$

$$\tan\frac{\pi}{4} = 1$$

$$\tan\frac{\pi}{2} = \infty$$

로부터

$$\tan^{-1} 0 = 0$$

$$\tan^{-1} 1 = \frac{\pi}{4}$$

$$\tan^{-1}(\infty) = \frac{\pi}{2}$$

이다.

이제 다음 적분을 생각하자.

$$I(a) = \int_0^\infty e^{-ax} \frac{\sin x}{x} dx \qquad (2\text{-}2\text{-}2)$$

이때 $a > 0$이다. 따라서 식 (2-2-1)의 좌변은 $I(0)$이다. 이 식에서 $\lim_{x \to \infty} e^{-ax} = 0$이므로

$$I(\infty) = 0$$

이다.

식 (2-2-2)를 a로 미분하면

$$\frac{dI}{da} = \int_0^\infty (-x) e^{-ax} \frac{\sin x}{x} dx$$

$$= -\int_0^\infty e^{-ax} \sin x \, dx \qquad (2\text{-}2\text{-}3)$$

가 된다. 여기서

$$J = \int_0^\infty e^{-ax} \sin x \, dx \qquad (2\text{-}2\text{-}4)$$

로 놓자. 부분적분법을 이용하면

$$J = \left[\left(-\frac{e^{-ax}}{a}\right) \sin x\right]_0^\infty - \int_0^\infty \left(-\frac{e^{-ax}}{a}\right) \cos x \, dx$$

또는

$$J = \frac{1}{a} \int_0^\infty e^{-ax} \cos x \, dx$$

가 된다. 한 번 더 부분적분법을 사용하면

$$J = \frac{1}{a}\left[\left(-\frac{e^{-ax}}{a}\right)\cos x\right]_0^\infty - \frac{1}{a}\int_0^\infty \left(-\frac{e^{-ax}}{a}\right)(-\sin x)\,dx$$

$$= \frac{1}{a^2} - \frac{1}{a^2}J$$

이므로

$$\left(1 + \frac{1}{a^2}\right)J = \frac{1}{a^2}$$

또는

$$J = \frac{1}{1+a^2}$$

이다. 이것을 식 (2-2-3)에 넣으면

$$\frac{dI}{da} = -\frac{1}{1+a^2}$$

이므로

$$I(a) = -\tan^{-1}a + C$$

로 쓸 수 있다. $I(\infty) = 0$이므로

$$0 = -\tan^{-1}(\infty) + C$$

가 되어,

$$C = \frac{\pi}{2}$$

이다. 따라서

$$I(a) = \frac{\pi}{2} - \tan^{-1} a$$

로부터

$$I(0) = \frac{\pi}{2}$$

임을 알 수 있다. 즉,

$$\int_0^\infty \frac{\sin x}{x} dx = \frac{\pi}{2}$$

가 성립한다. 이 식에서 $\frac{\sin x}{x}$ 는 우함수이므로

$$\int_{-\infty}^\infty \frac{\sin x}{x} dx = 2\int_0^\infty \frac{\sin x}{x} dx = \pi$$

이다.

디랙 델타 함수 _ 디랙이 만든 함수의 성질

정교수 이제 디랙이 새로 만든 재미있는 함수를 소개할 거야. 이것을 디랙 델타 함수라 부르고 $\delta(x)$로 쓰는데 다음과 같은 성질을 만족해.

$$\int_{-\infty}^{\infty} \delta(x)\,dx = 1 \qquad (2\text{-}3\text{-}1)$$

$$\int_{-\infty}^{\infty} \delta(x)f(x)\,dx = f(0) \qquad (2\text{-}3\text{-}2)$$

디랙 델타 함수는 디랙이 쓴 양자역학책 《양자역학의 원리(The Principles of Quantum Mechanics)》에 처음 등장했어.

물리군 처음 보는 함수예요.

정교수 예를 들어

$$\int_{-\infty}^{\infty} \delta(x)(3x^2 + 6x + 7)\,dx = 7$$

이 되지.

물리군 $f(x) = 3x^2 + 6x + 7$인 경우니까 $f(0) = 7$이군요.

정교수 맞아. 이번에는 다음 디랙 델타 함수를 봐.

$$\delta(x - x_0)$$

이 디랙 델타 함수는 다음 성질을 만족해.

$$\int_{-\infty}^{\infty} \delta(x - x_0) \, dx = 1 \qquad (2-3-3)$$

$$\int_{-\infty}^{\infty} \delta(x - x_0) f(x) \, dx = f(x_0) \qquad (2-3-4)$$

예를 들어

$$\int_{-\infty}^{\infty} \delta(x - 1)(3x^2 + 6x + 7) \, dx = 16$$

이 되지.

물리군 아주 신기한 함수네요.

정교수 여기서 잠깐! e^{Ax}을 미분하면 어떻게 되지?

물리군 Ae^{Ax}이에요.

정교수 그래. 이 성질을 이용하면

$$(e^{ikx})' = ike^{ikx}$$

이 돼. $e^{ikx} = f(x) + ig(x)$라고 하면

$$f' + ig' = ik(f + ig)$$

이고, 실수부와 허수부를 비교하면 다음과 같아.

$$f' = -kg$$

$$g' = kf$$

이 식을 만족하는 두 함수는

$$f = \cos kx$$

$$g = \sin kx$$

이므로

$$e^{ikx} = \cos kx + i \sin kx$$

이지. $k = 1$이면

$$e^{ix} = \cos x + i \sin x$$

가 돼. 이 공식을 잘 기억해 둬.

디랙 델타 함수는 다음과 같이 적분으로 나타낼 수 있어.

$$\delta(x) = \frac{1}{2\pi} \int_{-\infty}^{\infty} e^{ipx} dp \qquad (2-3-5)$$

$$\delta(x - x') = \frac{1}{2\pi} \int_{-\infty}^{\infty} e^{ip(x-x')} dp \qquad (2-3-6)$$

물리군 어떻게 디랙 델타 함수가 적분으로 표현되죠?

정교수 코시의 적분 공식을 이용하면 되는데, 차근차근 설명해 볼게.

일단

$$\delta_n(x) = \frac{\sin nx}{\pi x}$$

라고 놓자. 이제 우리는

$$\lim_{n \to \infty} \delta_n(x) = \delta(x) \qquad (2\text{-}3\text{-}7)$$

임을 보이려고 한다. 먼저

$$\int_{-\infty}^{\infty} \delta_n(x)\,dx = \int_{-\infty}^{\infty} \frac{\sin nx}{\pi x}\,dx$$

에서 $nx = t$로 치환하면

$$\int_{-\infty}^{\infty} \delta_n(x)\,dx = \frac{1}{\pi}\int_{-\infty}^{\infty} \frac{\sin t}{t}\,dt = 1$$

이다. 양변에 $\lim\limits_{n \to \infty}$ 를 취하면

$$\lim_{n \to \infty}\int_{-\infty}^{\infty} \delta_n(x)\,dx = 1$$

이 되어 식 (2-3-1)을 만족한다.

이번에는 다음 적분을 보자.

$$\int_{-\infty}^{\infty} \delta_n(x) F(x)\,dx \qquad (2\text{-}3\text{-}8)$$

여기서

$$F(x) = a_0 + a_1 x + a_2 x^2 + \cdots \tag{2-3-9}$$

으로 쓸 수 있다. 식 (2-3-8)에서 $nx = t$로 치환하면

$$\int_{-\infty}^{\infty} \delta_n(x) F(x)\,dx = \frac{1}{\pi} \int_{-\infty}^{\infty} \frac{\sin t}{t} F\left(\frac{t}{n}\right) dt \tag{2-3-10}$$

이다. 식 (2-3-9)를 식 (2-3-10)에 넣으면 다음과 같다.

$$\int_{-\infty}^{\infty} \delta_n(x) F(x)\,dx = \frac{1}{\pi} \int_{-\infty}^{\infty} \frac{\sin t}{t} \left[a_0 + a_1\left(\frac{t}{n}\right) + a_2\left(\frac{t}{n}\right)^2 + \cdots \right] dt$$

이 식의 양변에 $\lim_{n \to \infty}$를 취하면

$$\lim_{n \to \infty} \int_{-\infty}^{\infty} \delta_n(x) F(x)\,dx = \frac{a_0}{\pi} \int_{-\infty}^{\infty} \frac{\sin t}{t}\,dt = a_0 = F(0)$$

이 된다. $a_0 = F(0)$이므로 식 (2-3-2)가 성립한다. 따라서

$$\delta(x) = \lim_{n \to \infty} \frac{\sin nx}{\pi x} \tag{2-3-11}$$

이다. 한편 식 (2-3-11)로부터

$$e^{inx} = \cos nx + i \sin nx$$

$$e^{-inx} = \cos nx - i \sin nx$$

이고, 첫 번째 식에서 두 번째 식을 빼면

$$\sin nx = \frac{1}{2i}\left(e^{inx} - e^{-inx}\right)$$

이다. 따라서

$$\delta(x) = \lim_{n \to \infty} \frac{e^{inx} - e^{-inx}}{\pi 2ix}$$

$$= \frac{1}{2\pi} \lim_{n \to \infty} \frac{e^{inx} - e^{-inx}}{ix}$$

$$= \frac{1}{2\pi} \lim_{n \to \infty} \int_{-n}^{n} e^{ipx} dp$$

$$= \frac{1}{2\pi} \int_{-\infty}^{\infty} e^{ipx} dp$$

가 된다.

물리군   이해했어요.

하이젠베르크-보른-요르단-슈뢰딩거의 양자역학 _ 양자역학의 탄생 과정

정교수   여기서는 양자역학의 탄생 과정에 대해 조금 알아볼 거야.

양자역학은 1925년에서 1926년까지 하이젠베르크와 보른, 요르단, 슈뢰딩거가 구축했지.

간단하게 말해서 양자역학은 고전역학[3]을 양자화한다는 뜻이다. 고전역학에서는 물체의 위치를 알면 물체의 운동량[4]을 알 수 있다. 하지만 양자역학에서는 입자―이것을 양자라고 부른다―의 위치와 운동량을 동시에 정확하게 결정할 수 없다. 이것은

(위치의 불확정성) × (운동량의 불확정성)

이 0이 아님을 의미한다. 그러므로 위치와 운동량이 독립적이다.

고전역학에서 물체의 위치 x와 운동량 p는 수에 의해 묘사된다. 하지만 양자역학에서 물체(양자)의 위치와 운동량은 각각 위치 연산자 $\hat{x}$와 운동량 연산자 $\hat{p}$로 묘사되며 이 두 연산자는 다음 관계를 만족한다.

$$\hat{x}\hat{p} - \hat{p}\hat{x} = i\hbar \qquad (2\text{-}4\text{-}1)$$

여기서 $\hbar = \dfrac{h}{2\pi}$ 이고 h는 플랑크 상수라고 부르는 아주 작은 값이다. 위 식을 만족하는 경우는 두 가지이다. 하나는

---

3) 뉴턴 역학을 말한다.

4) 질량과 속도의 곱

$$\hat{x} = x$$

$$\hat{p} = \frac{\hbar}{i} \frac{\partial}{\partial x} \qquad (2\text{-}4\text{-}2)$$

이고 다른 하나는

$$\hat{x} = i\hbar \frac{\partial}{\partial p}$$

$$\hat{p} = p \qquad (2\text{-}4\text{-}3)$$

이다. 식 (2-4-2)를 위치 표현, 식 (2-4-3)을 운동량 표현이라고 부른다.

물리군 $\dfrac{\partial}{\partial x}$는 무슨 의미죠?

정교수 x에 대한 편미분을 말해. 즉, x만 문자로 생각하여 미분하는 걸 뜻하지. 그러니까

$$\frac{\partial}{\partial x}(p) = 0$$

이 돼. 일반적으로

$$\frac{\partial}{\partial x}(p\text{만의 함수}) = 0$$

이지. 마찬가지로 $\dfrac{\partial}{\partial p}$는 p에 대한 편미분이야. 즉, p만 문자로 생각하여 미분하는 걸 뜻해. 따라서

$$\frac{\partial}{\partial p}(x) = 0$$

이 돼. 일반적으로

$$\frac{\partial}{\partial p}(x\text{만의 함수}) = 0$$

이야.

양자역학에서는 전자가 주인공이다. 전자의 파동함수를 위치와 시간의 함수인 $\psi(x,t)$로 쓰거나 운동량과 시간의 함수인 $\phi(p,t)$로 쓸 수 있다. 이 파동함수에 대해 식 (2-4-2)는

$$x\frac{\hbar}{i}\frac{\partial}{\partial x}\psi(x,t) - \frac{\hbar}{i}\frac{\partial}{\partial x}(x\psi(x,t)) = i\hbar\psi(x,t)$$

가 되고, 식 (2-4-3)은

$$i\hbar\frac{\partial}{\partial p}(p\phi(p,t)) - pi\hbar\frac{\partial}{\partial p}\phi(p,t) = i\hbar\phi(p,t)$$

가 된다.

고전역학에서 역학적 에너지를 운동량과 위치의 함수로 나타낸 것을 해밀토니안 H라고 부른다. 즉,

$$H = \frac{1}{2m}p^2 + V(x)$$

이다. 여기서 m은 전자의 질량이고 V는 퍼텐셜에너지[5]이다.

양자역학에서는 해밀토니안도 해밀토니안 연산자 $\hat{H}$가 되어,

$$\hat{H} = \frac{1}{2m}(\hat{p})^2 + V(\hat{x}) \qquad (2\text{-}4\text{-}4)$$

이다. 해밀토니안 연산자는 시간에 대한 편미분으로 다음과 같이 표현한다.

$$\hat{H} \longrightarrow i\hbar \frac{\partial}{\partial t} \qquad (2\text{-}4\text{-}5)$$

그러므로 위치 표현으로 나타낸 슈뢰딩거 방정식은

$$i\hbar \frac{\partial}{\partial t} \psi(x,t) = \left[-\frac{\hbar^2}{2m} \frac{\partial^2}{\partial x^2} + V(x) \right] \psi(x,t) \qquad (2\text{-}4\text{-}6)$$

이고, 운동량 표현으로 나타낸 슈뢰딩거 방정식은

$$i\hbar \frac{\partial}{\partial t} \phi(p,t) = \left[\frac{p^2}{2m} + V\left(i\hbar \frac{\partial}{\partial p} \right) \right] \phi(p,t) \qquad (2\text{-}4\text{-}7)$$

이다.

물리군   두 가지 버전의 슈뢰딩거 방정식이 있군요.

정교수   맞아. 그런데 퍼텐셜에너지가 존재하는 것은 전자의 에너지

---

5) 위치에너지라고도 부른다.

가 일정한 걸 뜻해. 즉, 에너지가 시간에 따라 변하지 않는 거야. 이 에너지를 E라고 하면

$$\hat{H}\psi(x,t) = E\psi(x,t)$$

가 되지. 그러니까

$$i\hbar \frac{\partial}{\partial t}\psi(x,t) = E\psi(x,t) \quad (2\text{-}4\text{-}8)$$

라네. 이제

$$\psi(x,t) = \psi(x)T(t)$$

로 가정하고 이것을 식 (2-4-8)에 넣으면

$$i\hbar \frac{\partial}{\partial t}[\psi(x)T(t)] = E\psi(x)T(t)$$

이므로

$$\psi(x)i\hbar \frac{\partial}{\partial t}T(t) = E\psi(x)T(t)$$

가 돼. 그런데 T는 t만의 함수이니까 편미분 기호를 사용할 필요가 없어. 그러므로

$$i\hbar \frac{dT(t)}{dt} = ET(t)$$

또는

$$\frac{dT(t)}{dt} = -\frac{iE}{\hbar}T(t) \tag{2-4-9}$$

이지.

고등학교 수학에서 우리는 지수함수 e^x을 배우는데 일반적으로

$$\frac{d}{dt}e^{At} = Ae^{At}$$

이야. 이 식과 식 (2-4-9)를 비교하면

$$T(t) = e^{-\frac{iE}{\hbar}t} \tag{2-4-10}$$

으로 놓을 수 있어. 즉,

$$\psi(x,t) = e^{-\frac{iE}{\hbar}t}\psi(x) \tag{2-4-11}$$

이지. 이것을 위치 표현 슈뢰딩거 방정식에 넣으면 다음과 같아.

$$\left[-\frac{\hbar^2}{2m}\frac{\partial^2}{\partial x^2} + V(x)\right]\psi(x) = E\psi(x) \tag{2-4-12}$$

마찬가지로 운동량 표현을 이용하면

$$\phi(p,t) = e^{-\frac{iE}{\hbar}t}\phi(p) \tag{2-4-13}$$

이고, 이것을 운동량 표현 슈뢰딩거 방정식에 넣으면

$$\left[\frac{p^2}{2m} + V\left(i\hbar\frac{\partial}{\partial p}\right)\right]\phi(p) = E\phi(p) \qquad (2\text{-}4\text{-}14)$$

로 쓸 수 있지.

보른은 $|\psi(x,t)|^2$이 시각 t, 위치 x에서 전자를 발견할 확률인 것을 알아냈어. 이 확률은 시간에 따라 달라지지 않아.

$$|\psi(x,t)|^2 = |\psi(x)|^2$$

그러므로

$$\int_{-\infty}^{\infty} |\psi(x,t)|^2 dx = \int_{-\infty}^{\infty} |\psi(x)|^2 dx = 1$$

이야.

마찬가지로 $|\phi(p,t)|^2$은 시각 t일 때 운동량 p인 전자를 발견할 확률이고, 이 확률은 시간에 따라 달라지지 않아.

$$|\phi(p,t)|^2 = |\phi(p)|^2$$

따라서

$$\int_{-\infty}^{\infty} |\phi(p,t)|^2 dp = \int_{-\infty}^{\infty} |\phi(p)|^2 dp = 1$$

이 되지.

양자역학의 위치 상태와 운동량 상태 _ 전자의 위치와 운동량

정교수   디랙은 양자역학의 위치 연산자와 운동량 연산자에 대응하는 위치 상태와 운동량 상태를 켓벡터로 나타내고 싶어 했어. 양자역학은 전자가 주인공이니까 전자의 위치와 전자의 운동량이 되겠지. 그 내용을 자세히 알아볼게.

일차원에서 전자가 가질 수 있는 위치가

$x_1, x_2, x_3, \cdots$

으로 주어진다고 하자. 각각의 위치에 대응하는 양자 상태를 다음과 같이 켓벡터로 나타내자.

$|x_1\rangle, |x_2\rangle, |x_3\rangle, \cdots$

이 켓벡터에 위치 연산자 $\hat{x}$를 작용하면 위치가 고윳값으로 나와야 하므로

$\hat{x}|x_1\rangle = x_1|x_1\rangle$

$\hat{x}|x_2\rangle = x_2|x_2\rangle$

$\hat{x}|x_3\rangle = x_3|x_3\rangle$

$\vdots$

이 된다. 여기서 켓벡터는 정규직교화한 벡터이다.

전자의 허용된 위치가 다음 두 경우인 때를 예로 들어 좀 더 쉽게 설명하겠다.

$x_1 = 1$

$x_2 = 2$

이 두 위치에 대응하는 위치 켓벡터를 $|x_1\rangle, |x_2\rangle$라고 하면

$$\hat{x}|x_1\rangle = x_1|x_1\rangle \tag{2-5-1}$$

$$\hat{x}|x_2\rangle = x_2|x_2\rangle \tag{2-5-2}$$

이다.

이때 두 켓벡터는 정규직교화한 벡터이므로 다음과 같이 행렬로 나타낼 수 있다.

$$|x_1\rangle = \begin{pmatrix} 1 \\ 0 \end{pmatrix}$$

$$|x_2\rangle = \begin{pmatrix} 0 \\ 1 \end{pmatrix}$$

두 켓벡터가 정규직교화된 것은 다음을 통해 알 수 있다.

$$\langle x_1 | x_1 \rangle = 1, \quad \langle x_2 | x_2 \rangle = 1$$

$$\langle x_1 | x_2 \rangle = \langle x_2 | x_1 \rangle = 0$$

$$| x_1 \rangle\langle x_1 | + | x_2 \rangle\langle x_2 | = I$$

물리군 $\hat{x}$도 행렬로 나타낼 수 있나요?

정교수 물론이야.

$$\hat{x} = \begin{pmatrix} x_1 & 0 \\ 0 & x_2 \end{pmatrix}$$

로 나타낼 수 있지.

물리군 확인해 볼게요.

$$\begin{pmatrix} x_1 & 0 \\ 0 & x_2 \end{pmatrix}\begin{pmatrix} 1 \\ 0 \end{pmatrix} = \begin{pmatrix} x_1 \\ 0 \end{pmatrix} = x_1 \begin{pmatrix} 1 \\ 0 \end{pmatrix}$$

이니까 식 (2-5-1)이 성립하고

$$\begin{pmatrix} x_1 & 0 \\ 0 & x_2 \end{pmatrix}\begin{pmatrix} 0 \\ 1 \end{pmatrix} = \begin{pmatrix} 0 \\ x_2 \end{pmatrix} = x_2 \begin{pmatrix} 0 \\ 1 \end{pmatrix}$$

이니까 식 (2-5-2)가 성립해요.

정교수 그럼 $\hat{x}$의 전치를 구해 봐.

물리군 다음과 같아요.

$$\begin{pmatrix} x_1 & 0 \\ 0 & x_2 \end{pmatrix}^T = \begin{pmatrix} x_1 & 0 \\ 0 & x_2 \end{pmatrix}$$

그러니까

$$\hat{x}^T = \hat{x} \qquad (2\text{-}5\text{-}3)$$

가 돼요.

정교수 그렇지. 위치 연산자에 전치를 취해도 달라지지 않아.

임의의 상태 $|\psi\rangle$는 전자가 가질 수 있는 두 위치 상태 켓벡터의 중첩으로

$$|\psi\rangle = \psi_1 |x_1\rangle + \psi_2 |x_2\rangle$$

가 된다. 이 식의 양변에 $\langle x_1|$을 곱하면

$$\psi_1 = \langle x_1 | \psi \rangle$$

이고, 양변에 $\langle x_2|$를 곱하면

$$\psi_2 = \langle x_2 | \psi \rangle$$

이다. 이때

$$\langle \psi | \psi \rangle = \langle \psi | (|x_1\rangle\langle x_1| + |x_2\rangle\langle x_2|) | \psi \rangle$$

$$= \langle \psi | x_1 \rangle\langle x_1 | \psi \rangle + \langle \psi | x_2 \rangle\langle x_2 | \psi \rangle$$

$$= |\langle x_1 | \psi \rangle|^2 + |\langle x_2 | \psi \rangle|^2$$

$$= |\psi_1|^2 + |\psi_2|^2$$

$$= 1$$

이 된다. 여기서 $|\psi_1|^2 = |\langle x_1 | \psi \rangle|^2$은 전자를 위치 x_1에서 발견할 확률이고 $|\psi_2|^2 = |\langle x_2 | \psi \rangle|^2$은 전자를 위치 x_2에서 발견할 확률이다.

예를 들어

$$|\psi\rangle = \frac{1}{5}\begin{pmatrix} 3 \\ 4 \end{pmatrix}$$

인 경우를 생각하자. 이것은

$$\langle \psi | \psi \rangle = 1$$

을 만족한다. $|\psi\rangle$는 두 위치 상태의 중첩으로

$$|\psi\rangle = \psi_1 |x_1\rangle + \psi_2 |x_2\rangle$$

로 나타낼 수 있다. 즉,

$$\psi_1 = \frac{3}{5}$$

$$\psi_2 = \frac{4}{5}$$

이다. 그러므로 상태 $|\psi\rangle$일 때 전자를 위치 x_1에서 발견할 확률은

$$|\psi_1|^2 = \frac{9}{25}$$

이고, 상태 $|\psi\rangle$일 때 전자를 위치 x_2에서 발견할 확률은

$$|\psi_2|^2 = \frac{16}{25}$$

이다.

물리군 전자가 가질 수 있는 위치가 두 개보다 많으면 어떻게 되나요?

정교수 만일 전자가 가질 수 있는 위치가 $x_1, x_2, x_3, \cdots$이라면

$$\hat{x}|x_i\rangle = x_i|x_i\rangle \qquad (i = 1, 2, 3, \cdots) \tag{2-5-4}$$

로 쓸 수 있고 켓벡터들은

$$\sum_{k=1}^{\infty}|x_k\rangle\langle x_k| = I \tag{2-5-5}$$

를 만족한다. 임의의 상태를 $|\psi\rangle$라고 하면

$$|\psi\rangle = \sum_{k=1}^{\infty} \psi_k |x_k\rangle \qquad (2\text{-}5\text{-}6)$$

가 된다. 이때

$$\psi_k = \langle x_k | \psi \rangle \qquad (2\text{-}5\text{-}7)$$

이고, 임의의 상태 $|\psi\rangle$는 전자들이 있을 수 있는 가능한 위치 상태 켓벡터의 중첩으로 다음과 같이 쓸 수 있다.

$$\begin{aligned}|\psi\rangle &= \sum_{k=1}^{\infty} \langle x_k | \psi \rangle | x_k \rangle \\ &= \sum_{k=1}^{\infty} |x_k\rangle \langle x_k | \psi \rangle \end{aligned} \qquad (2\text{-}5\text{-}8)$$

이때 상태 $|\psi\rangle$에서 위치가 x_k인 전자를 발견할 확률은

$$|\psi_k|^2 = |\langle x_k | \psi \rangle|^2$$

이다.

물리군 하지만 일차원에서 전자의 위치는 연속적으로 변하잖아요?
정교수 그래. 일차원에서 전자의 위치는 실수가 되지. 예를 들면 다음과 같아.

$$\hat{x}\,|\,\sqrt{2}\,\rangle = \sqrt{2}\,|\,\sqrt{2}\,\rangle$$

$$\hat{x}\,|\,0.17\,\rangle = 0.17\,|\,0.17\,\rangle$$

실수로 표현되는 임의의 위치를 x라 하고 그 위치에 대응하는 양자상태를 $|\,x\,\rangle$라고 하면 식 (2-5-4)처럼

$$\hat{x}\,|\,x\,\rangle = x\,|\,x\,\rangle \tag{2-5-9}$$

로 쓸 수 있지. 여기서 x는 모든 실수를 의미해. 이렇게 모든 실수에 대해 더하는 것은 적분으로 표현되니까 식 (2-5-5)는 다음과 같이 나타낼 수 있지.

$$\int_{-\infty}^{\infty} |\,x\,\rangle\langle\,x\,|\,dx = I \tag{2-5-10}$$

이때 임의의 상태를 뜻하는 켓벡터 $|\,\psi\,\rangle$는 전자가 있을 수 있는 모든 위치 상태 켓벡터의 중첩이 돼. 그러니까 다음과 같아.

$$|\,\psi\,\rangle = \int \psi(x)\,|\,x\,\rangle\,dx \tag{2-5-11}$$

여기서 $\psi(x)$를 위치 공간에서의 파동함수라고 불러.

물리군 식 (2-5-6)에서 합의 기호가 적분으로 바뀐 거군요.
정교수 맞아. 앞으로 $\int_{-\infty}^{\infty}$를 간단하게 $\int$로 쓸게. 이제 식 (2-5-7)은 다음과 같이 나타낼 수 있지.

$$\langle x | \psi \rangle = \psi(x) \tag{2-5-12}$$

물리군 왜 그런가요?

정교수 위치 x에 대응하는 켓벡터를 $|x\rangle$, 위치 x'에 대응하는 켓벡터를 $|x'\rangle$이라고 해 봐. 이때

$$\hat{x}|x\rangle = x|x\rangle \tag{2-5-13}$$

$$\hat{x}|x'\rangle = x'|x'\rangle \tag{2-5-14}$$

으로 쓸 수 있지. 식 (2-5-13)의 양변 왼쪽에 $\langle x'|$을 곱하면

$$\langle x'|\hat{x}|x\rangle = x\langle x'|x\rangle \tag{2-5-15}$$

가 돼. 식 (2-5-14)의 전치를 취하면

$$(\hat{x}|x'\rangle)^T = (|x'\rangle)^T (\hat{x})^T = \langle x'|\hat{x}$$

이므로

$$\langle x'|\hat{x} = x'\langle x'|$$

이야. 이 식의 양변 오른쪽에 $|x\rangle$를 곱하면

$$\langle x'|\hat{x}|x\rangle = x'\langle x'|x\rangle \tag{2-5-16}$$

이지. 식 (2-5-15)에서 식 (2-5-16)을 빼면

$$0 = (x-x')\langle x'|x\rangle \tag{2-5-17}$$

가 돼. 두 위치가 다르면, 즉 $x \neq x'$이면

$$\langle x'|x\rangle = 0$$

이어야 하고, $x = x'$이면 $\langle x'|x\rangle$가 0일 필요가 없어. 이런 성질을 가진 함수가 바로 디랙 델타 함수야. 그러니까

$$\langle x'|x\rangle = \langle x|x'\rangle = \delta(x-x') \tag{2-5-18}$$

으로 놓을 수 있어. 그러므로 다음과 같아.

$$\langle x|\psi\rangle = \langle x|\int \psi(x')|x'\rangle dx'$$

$$= \int \psi(x')\langle x|x'\rangle dx'$$

$$= \int \psi(x')\delta(x-x')dx'$$

$$= \psi(x)$$

그리고 식 (2-5-11)은

$$|\psi\rangle = I|\psi\rangle = \int |x\rangle\langle x|\psi\rangle dx \tag{2-5-19}$$

로 쓸 수도 있어.

물리군 식 (2-5-11)에 나오는 $|\psi\rangle$의 크기는 얼마예요?

정교수 그건 $\langle\psi|\psi\rangle$를 계산해 보면 돼. 이 값에 루트를 취하면 $|\psi\rangle$의 크기이므로

$$\langle\psi|\psi\rangle = \langle\psi|\int|x\rangle\langle x|\psi\rangle dx$$
$$= \int \langle\psi|x\rangle\langle x|\psi\rangle dx$$

가 돼. 그런데 양자역학에서는 $|\psi\rangle$가 복소수일 수도 있으니까

$$\langle\psi|x\rangle = \langle x|\psi\rangle^*$$

이지. 따라서

$$\langle\psi|\psi\rangle = \int \langle x|\psi\rangle^* \langle x|\psi\rangle dx$$
$$= \int \psi(x)^* \psi(x) dx$$
$$= \int |\psi(x)|^2 dx$$

라네. $|\psi(x)|^2$은 전자를 위치 x에서 발견할 확률이고 확률을 모두 더하면 1이므로 다음과 같아.

$$\langle\psi|\psi\rangle = \int |\psi(x)|^2 dx = 1$$

즉, 식 (2-5-11)에 나오는 $|\psi\rangle$의 크기는 1이야.

물리군 그렇군요. 양자역학에서는 위치 연산자뿐 아니라 운동량 연

산자도 나타나잖아요? 운동량 연산자에 대응하는 켓벡터도 있나요?

정교수 물론이야. 운동량 p에 대응하는 켓벡터를 $|p\rangle$라고 쓰는데

$$\hat{p}|p\rangle = p|p\rangle \qquad (2\text{-}5\text{-}20)$$

라네. 여기서 운동량 p는 모든 실숫값을 갖지. 이때

$$\int |p\rangle\langle p|\,dp = I \qquad (2\text{-}5\text{-}21)$$

가 돼. $\int$은 $\int_{-\infty}^{\infty}$를 줄여 쓴 거야. 위치 연산자의 경우와 마찬가지로 운동량 연산자에 대한 켓벡터는 다음 성질을 만족해.

$$\langle p|p'\rangle = \delta(p - p') \qquad (2\text{-}5\text{-}22)$$

이제 $|p\rangle$를 기저로 갖는 벡터 공간을 운동량 공간이라고 부를 거야. 앞에서 우리는 임의의 상태 $|\psi\rangle$를 생각했어. 이 상태를 운동량 공간에서의 기저로 나타내면

$$|\psi\rangle = \int \phi(p)|p\rangle\,dp \qquad (2\text{-}5\text{-}23)$$

이고, $\phi(p)$를 운동량 공간에서의 파동함수라고 불러. 여기서

$$\phi(p) = \langle p|\psi\rangle \qquad (2\text{-}5\text{-}24)$$

가 돼. 그러니까 임의의 상태 $|\psi\rangle$에 $\langle x|$를 내적시키면 위치 공간의 파동함수를 얻고, $\langle p|$를 내적시키면 운동량 공간에서의 파동함수를

얻지.

물리군 파동함수를 위치 공간에서 묘사할 수도 있고 운동량 공간에서 묘사할 수도 있네요.

정교수 맞아. 위치 공간에서는 $\psi(x)$가 되고 운동량 공간에서는 $\phi(p)$가 되는 거야.

물리군 위치 공간의 켓벡터와 운동량 공간의 켓벡터 사이에 어떤 관계가 있나요?

정교수 이제 그걸 생각해 보려고 해. 식 (2-5-21)을 이용하면

$$\int \langle x | p \rangle \langle p | x' \rangle dp = \delta(x-x')$$

이야. 우리는 디랙 델타 함수에서

$$\delta(x-x') = \frac{1}{2\pi} \int e^{ia(x-x')} da$$

인 것을 배웠어. 이 식에서

$$a = \frac{p}{\hbar}$$

라고 두면

$$\delta(x-x') = \frac{1}{2\pi\hbar} \int e^{\frac{i}{\hbar}p(x-x')} dp$$

이지. 그러니까

$$\int \langle x | p \rangle \langle p | x' \rangle dp = \frac{1}{2\pi\hbar} \int e^{\frac{i}{\hbar}p(x-x')} dp$$

$$= \int \left(\frac{1}{\sqrt{2\pi\hbar}} e^{-\frac{i}{\hbar}px}\right)^* \left(\frac{1}{\sqrt{2\pi\hbar}} e^{-\frac{i}{\hbar}px'}\right) dp$$

가 돼. 따라서

$$\langle p | x \rangle = \frac{1}{\sqrt{2\pi\hbar}} e^{-\frac{i}{\hbar}px}$$

$$\langle x | p \rangle = \langle p | x \rangle^* = \frac{1}{\sqrt{2\pi\hbar}} e^{\frac{i}{\hbar}px} \tag{2-5-25}$$

인 것을 알 수 있다네.

물리군 위치 공간에서의 파동함수와 운동량 공간에서의 파동함수는 어떤 관계가 있죠?

정교수 두 파동함수 사이에는 다음과 같은 관계가 있어.

$$\phi(p) = \frac{1}{\sqrt{2\pi\hbar}} \int \psi(x) e^{-\frac{i}{\hbar}px} dx \tag{2-5-26}$$

$$\psi(x) = \frac{1}{\sqrt{2\pi\hbar}} \int \phi(p) e^{\frac{i}{\hbar}px} dp \tag{2-5-27}$$

물리군 이건 어떻게 증명해요?

정교수 식 (2-5-26)만 증명해 줄게.

$$\phi(p) = \langle p | \psi \rangle$$
$$= \int \langle p | x \rangle \langle x | \psi \rangle dx$$
$$= \int \langle p | x \rangle \psi(x) dx$$
$$= \frac{1}{\sqrt{2\pi\hbar}} \int \psi(x) e^{-\frac{i}{\hbar}px} dx$$

물리군 브라켓 기호를 이용해 슈뢰딩거 방정식을 만들 수 있나요?
정교수 물론이야. 먼저 다음 식을 봐.

$$\langle x | \hat{p} | x' \rangle = \int \langle x | \hat{p} | p \rangle \langle p | x' \rangle dp$$
$$= \int p \langle x | p \rangle \langle p | x' \rangle dp$$

여기서

$$p \langle x | p \rangle = p \frac{1}{\sqrt{2\pi\hbar}} e^{\frac{i}{\hbar}px}$$
$$= \frac{\hbar}{i} \frac{\partial}{\partial x} \frac{1}{\sqrt{2\pi\hbar}} e^{\frac{i}{\hbar}px}$$
$$= \frac{\hbar}{i} \frac{\partial}{\partial x} \langle x | p \rangle$$

이므로

$$\langle x|\hat{p}|x'\rangle = \frac{\hbar}{i}\frac{\partial}{\partial x}\int \langle x|p\rangle\langle p|x'\rangle dp$$

$$= \frac{\hbar}{i}\frac{\partial}{\partial x}\delta(x-x')$$

이 돼. 마찬가지로

$$\langle x|\hat{p}^2|x'\rangle = \left(\frac{\hbar}{i}\frac{\partial}{\partial x}\right)^2 \delta(x-x')$$

이지. 이제 $|\psi\rangle$가 만족하는 슈뢰딩거 방정식

$$\left[\frac{1}{2m}\hat{p}^2 + V(\hat{x})\right]|\psi\rangle = E|\psi\rangle$$

를 생각해 봐. 이 식 양변의 왼쪽에서 $\langle x|$를 가하면

$$\langle x|\left[\frac{1}{2m}\hat{p}^2 + V(\hat{x})\right]|\psi\rangle = E\langle x|\psi\rangle$$

이고, 좌변은

$$\langle x|\left[\frac{1}{2m}\hat{p}^2 + V(\hat{x})\right]|\psi\rangle$$

$$= \int \left[\frac{1}{2m}\langle x|\hat{p}^2|x'\rangle\langle x'|\psi\rangle + \langle x|V(\hat{x})|x'\rangle\langle x'|\psi\rangle\right]dx'$$

이 되지. 여기서

$$\int \langle x \mid V(\hat{x}) \mid x' \rangle \langle x' \mid \psi \rangle dx'$$

$$= \int V(x') \langle x \mid x' \rangle \langle x' \mid \psi \rangle dx'$$

$$= \int V(x') \delta(x-x') \psi(x') dx'$$

$$= V(x) \psi(x)$$

이고

$$\int \langle x \mid \hat{p}^2 \mid x' \rangle \langle x' \mid \psi \rangle dx'$$

$$= -\hbar^2 \int \left(\frac{\partial^2}{\partial x^2} \delta(x-x') \right) \psi(x') dx'$$

$$= -\hbar^2 \int \left(\frac{\partial^2}{\partial x'^2} \delta(x-x') \right) \psi(x') dx'$$

인데 부분적분을 하면 다음과 같아.

$$\int \langle x \mid \hat{p}^2 \mid x' \rangle \langle x' \mid \psi \rangle dx'$$

$$= -\hbar^2 \int \delta(x-x') \frac{\partial^2}{\partial x'^2} \psi(x') dx'$$

$$= -\hbar^2 \frac{\partial^2}{\partial x^2} \psi(x)$$

따라서 슈뢰딩거 방정식

$$-\frac{\hbar^2}{2m}\frac{\partial^2}{\partial x^2}\psi(x)+V(x)\psi(x)=E\psi(x)$$

가 성립하는 것을 알 수 있어.

물리군   그렇군요.

세 번째 만남

파인먼의 경로 적분

파인먼의 생애 _ 수학 천재의 사랑과 도전

정교수 파인먼의 경로 적분을 설명하기에 앞서 양자전기역학의 창시자 중 한 명이자 천재 물리학자인 파인먼을 소개하겠네.

파인먼(Richard Phillips Feynman, 1918~1988, 1965년 노벨 물리학상 수상)

파인먼은 1918년 5월 11일 미국 뉴욕 퀸스에서 태어났다. 그의 아버지는 벨라루스 민스크(당시 러시아 제국의 일부)의 유대인 가정에서 태어나 다섯 살 때 부모와 함께 미국으로 이민했다. 파인먼의 어머니는 미국의 유대인 가정에서 태어났다.

파인먼이 다섯 살이었을 때, 남동생 헨리 필립스가 태어났지만 4주 만에 사망했다. 그로부터 4년 후 여동생 조앤이 태어났고 가족은 퀸스의 파로커웨이로 이사했다. 조앤은 훗날 천체물리학자가 되었다.

어릴 때부터 기존 사고방식에 도전하는 것을 즐긴 파인먼은 유머 감각 또한 출중했다. 그는 공학에 재능이 있어 집에 실험실을 만들었으며 라디오 수리하기를 좋아했다. 초등학교에 다닐 때는 집에 도난 경보 시스템을 만들기도 했다.

뉴욕 파로커웨이 고등학교에 들어가자마자, 파인먼은 빠르게 수학 상급반으로 올라갔다. 그는 15세 때 삼각법, 고급 대수학, 무한급수, 해석기하학, 미분적분학을 독학으로 공부할 정도로 수학을 잘했다. 고등학교 마지막 해에는 뉴욕 대학 수학 선수권 대회에서 우승했다.

파인먼은 컬럼비아 대학에 지원했지만 유대인 학생 수 할당제 때문에 받아들여지지 않았다. 대신 매사추세츠 공과대학(MIT)에 들어갔고 그곳에서 파이 람다 피(Pi Lambda Phi) 남학생 클럽에 참여했다.

매사추세츠 공과대학

원래 수학을 전공한 파인먼은 수학이 너무 추상적이라고 생각하여 나중에 전공을 전기공학으로, 그러다 다시 물리학으로 바꾸었다. 대학 시절 그는 세계적인 학술지인 《피지컬 리뷰》에 논문 두 편을 발표했다. 첫 번째 논문은 지도교수인 바야르타와 함께 쓴 것으로 은하의

별들에 의한 우주선의 산란 연구였다. 두 번째 논문은 첫 단독 논문으로 분자의 힘에 관한 연구였다.

1939년에 파인먼은 학사 학위를 받았고 프린스턴 대학 물리학 대학원 입학시험을 쳤다. 이 시험에서 그는 수학에서 뛰어난 점수를 받아 합격했다. 1942년에는 프린스턴 대학에서 박사 학위를 받았는데 그의 논문 지도교수는 존 아치볼드 휠러였다. 파인먼은 〈양자역학에서 최소 작용의 원리〉라는 제목의 논문으로 박사 학위를 받았다. 여기서 그는 최소 작용의 원리를 양자역학 문제에 적용했고 경로 적분 공식화와 파인먼 다이어그램의 개념을 만들었다.

파인먼은 고등학교 때 연인인 알린 그린바움(Arline Greenbaum)이 결핵으로 심각하게 아프다는 사실에도 불구하고 박사 학위를 받으면 그녀와 결혼하기로 결심했다. 그녀의 병은 당시 불치병이었고, 의사들은 2년 이상 살지 못할 거라고 말했다. 하지만 1942년 6월 29일, 두 사람은 페리호를 타고 스태튼아일랜드로 갔고 그곳에서 결혼했다. 식이 끝난 후 파인먼은 그린바움을 데버라 병원으로 데려갔고, 주말마다 그녀를 방문했다.

유럽에서는 제2차 세계대전이 발발했지만 미국은 아직 전쟁이 시작되지 않은 1941년 여름, 파인먼은 펜실베이니아의 프랭크퍼드 무기고에서 탄도 문제를 연구하며 지냈다. 진주만 공격이 미국을 전쟁으로 몰아넣은 후, 파인먼은 프린스턴 대학의 윌슨이 이끄는 팀에 들어갔다. 이 팀은 맨해튼 프로젝트의 일환으로 원자폭탄에 사용할 농축 우라늄을 생산하는 방법을 연구했는데, 우라늄-235와 우라늄-238을 전자기적으로 분리하기 위한 아이소트론(isotron)이라는 장치를 개발하고 있었다.

1943년 초, 로버트 오펜하이머는 뉴멕시코의 메사에 원자폭탄을 설계하고 제조할 비밀 장소인 로스앨러모스 연구소를 설립하고 있었다. 그리고 프린스턴의 윌슨 팀에게 그곳으로 이동하자고 제안했다. 윌슨은 이를 받아들였고, 파인먼은 아내와 함께 1943년 3월 28일에 기차를 타고 뉴멕시코로 갔다. 철도 회사는 파인먼의 아내를 위해 휠체어를 제공했고, 파인먼은 그녀를 위해 로스앨러모스에서 개인적으로 방을 구했다.

파인먼의 로스앨러모스 신분증 사진
(출처: 로스앨러모스 국립연구소)

로스앨러모스에서 파인먼은 한스 베테가 주도하는 이론 부서에 소속되었다. 그와 베테는 핵분열 폭탄의 폭발력을 계산하기 위한 베테-파인먼 방정식을 개발했다. 또한 파인먼은 핵분열성 물질 조립체가 임계치에 얼마나 가까운지를 측정하기 위해 소형 원자로 로스앨러모스 '물 보일러'의 중성자 방정식을 계산했다.

이 작업을 완료한 후 파인먼은 맨해튼 프로젝트가 우라늄 농축 시설을 건설했던 테네시주 오크리지에 있는 클린턴 엔지니어 제작소로 파견되었다. 그는 특히 농축 우라늄이 중성자 감속재 역할을 하는 물과 접촉할 때 임계 사고를 피할 수 있도록 재료 저장을 위한 안전 절차를 고안했다. 그는 농축되지 않은 우라늄은 얼마든지 안전하게 저장할 수 있지만 농축 우라늄은 조심스럽게 다루어야 한다고 주장했고, 다양한 등급의 농축에 대한 일련의 안전 권장 사항을 개발했다.

테이블 가운데가 파인먼
(출처: 미국 물리학 연구소)

로스앨러모스로 돌아온 파인먼은 제안된 수소화 우라늄 폭탄에 대한 이론적인 작업과 계산을 하는 그룹의 책임자가 되었다. 보안을 위해 격리된 로스앨러모스에서 파인먼은 동료들의 캐비닛과 책상에 있는 잠금장치를 푸는 장난을 쳤다. 그는 물리학자들이 사용하는 비밀

번호가 수학이나 물리학 상수와 관련된다는 사실을 가지고 이들의 캐비닛을 쉽게 열 수 있었다. 예를 들어 어떤 물리학자는 오일러 수 $e = 2.71828\cdots$로부터 비밀번호를 27-18-28로 해두었고, 파인먼은 이 캐비닛의 비밀번호를 추측해 열었다.

로스앨러모스에서 받는 파인먼의 월급 380달러는 생활비와 아내의 의료비 지출에 필요한 금액의 절반에 지나지 않았다. 그들은 알린이 저축한 돈 3,300달러를 쓸 수밖에 없었다.

맨해튼 프로젝트 당시 파인먼(가운데)

파인먼의 보살핌에도 불구하고 아내 알린은 1945년 6월 16일에 사망했다. 그 후 파인먼은 다시 프로젝트 작업에 몰두했고 트리니티 핵실험에 참관했다. 파인먼은 어두운 안경이나 용접용 렌즈 없이 폭발을 목격했다. 폭발의 엄청난 밝기로 그는 트럭 바닥에 몸을 숨겨야 했다.

전쟁이 끝난 후 파인먼은 코넬 대학에서 물리학을 가르쳤다. 그러던 어느 날 식당에서 누군가가 저녁 식사 접시를 공중에 던진 사건에서 영감을 받아 빙빙 돌면서 운동하는 원판의 물리학을 떠올렸다. 그는 이것을 전자의 상대론적 이론에 적용하고자 했다. 파인먼은 코넬 대학에서 양자전기역학을 연구하고 있었으며 이 결과를 1949년에 논문으로 발표했다. 이 논문으로 파인먼은 양자전기역학의 창시자 중 한 명이 되었다.

1949년 7월, 파인먼은 브라질 리우데자네이루에서 몇 주를 보냈다. 그해에 소련(현재 러시아를 포함한 여러 나라의 연방)이 첫 번째 원자폭탄을 터뜨리자 과학자 스파이 문제가 대두되었다. 당시 많은 물리학자가 스파이로 의심받았다. 물리학자 데이비드 봄은 1950년 12월 4일에 체포되었고, 이듬해 10월에 브라질로 이주했다.

파인먼은 1951년부터 1952년에 안식년을 맞이하여 이 기간을 브라질에서 보냈다. 그는 삼바 음악에 깊은 인상을 받았고 콩가 드럼을 연주하기도 했다.

안식년이 끝난 후 파인먼은 코넬 대학에 돌아가지 않고 캘리포니아 공과대학(칼텍, Caltech)으로 갔다. 그는 칼텍 교수로 있는 동안 두 번째 결혼을 했다. 그의 두

번째 부인은 미국 캔자스주 니오데샤 출신의 메리 루이즈 벨(Mary Louise Bell)이었다.

두 사람은 코넬 대학의 한 카페테리아에서 만났다. 메리는 코넬 대학에서 멕시코 미술과 직물의 역사를 공부했는데, 나중에 파인먼을 따라 칼텍으로 가서 강의를 했다. 파인먼이 브라질에 있는 동안 그녀는 미시간 주립 대학에서 가구 및 인테리어의 역사를 가르쳤다.

파인먼은 리우데자네이루에서 편지로 메리에게 청혼했으며, 두 사람은 파인먼이 미국으로 돌아온 후인 1952년 6월 28일에 아이다호주 보이시에서 결혼했다. 둘은 정치 성향이 달라 자주 다투었고, 1956년 5월 20일에 별거를 시작해 1958년 5월 5일에 이혼했다.

칼텍에서 파인먼은 과냉각된 액체 헬륨의 초유체 물리학을 연구했다. 초유체란 점성이 거의 사라진 유체를 말한다. 파인먼은 소련 물리학자 란다우의 초유체 이론에 대한 양자역학적 설명을 제공했다. 또한 그는 쿼크로 노벨 물리학상을 수상한 머리 겔만과 함께 약한 붕괴의 모형을 개발했다. 그리고 핵자 산란을 지배하는 강한 상호작용에 대해 쪽입자(parton) 모형을 만들었다.

1960년대 초 파인먼은 칼텍의 학부생들을 위한 물리학 강의를 했다. 이 수업 내용은 훗날 《파인먼의 물

리학 강의》라는 책으로 출간되었다.

 1960년대에 파인먼은 자서전을 쓰기로 결심하고, 역사학자들에게 인터뷰를 허락했다. 1980년대에 그는 랠프 레이턴(로버트 레이턴의 아들)과 함께 일하면서 테이프 리코더에 여러 시기의 일을 녹음했고, 이것을 랠프가 글로 옮겼다. 이 내용은 1985년에 《파인먼 씨, 농담도 잘하시네!》라는 제목으로 출판되어 베스트셀러에 올랐다.

 1978년에 파인먼은 복통으로 치료를 하다가 희귀암인 지방육종 진단을 받았다. 외과 의사들은 한쪽 신장과 비장을 억누른 축구공 크기의 종양을 제거했다. 하지만 1986년 10월과 1987년 10월의 추가 수술 후 파인먼은 1988년 2월 15일 69세의 나이로 사망했다.

파인먼의 아이디어 _ 양자역학을 새롭게 이해하는 이론

정교수 파인먼이 대학교 3학년 양자역학 수업 시간에 발견한 경로 적분 이야기를 하려고 해. 남들과 다르게 사고하는 그는 양자역학 수업 도중에 참신한 아이디어를 떠올렸어. 그것은 양자역학을 새롭게 이해하는 이론이 되었지. 훗날 그는 이 내용을 보강해 박사 학위 논문으로 만들었다네.

물리군 어떤 내용이죠?

정교수 파인먼은 디랙의 브라켓 기호를 좋아했어. 그래서 디랙처럼 전자를 나타내는 양자 상태 $|\psi\rangle$를 생각했지. 파인먼은 해밀토니안이 시간의 함수인 경우를 고려했어. 그 경우 파동함수는 시간과 위치의 함수가 돼. 그러니까

$$\psi(x) = \langle x | \psi \rangle$$

처럼

$$\psi(x, t) = \langle x | \psi(t) \rangle \qquad (3\text{-}2\text{-}1)$$

또는

$$\psi(x, t) = \langle x, t | \psi \rangle \qquad (3\text{-}2\text{-}2)$$

로 쓰기로 했어.

물리군 그렇다면 $|\psi(t)\rangle$는 시간 의존형 슈뢰딩거 방정식을 만족하

겠군요.

정교수 맞아. $|\psi(t)\rangle$는

$$i\hbar \frac{\partial}{\partial t} |\psi(t)\rangle = H |\psi(t)\rangle \tag{3-2-3}$$

를 만족하지. 그런데 $|\psi(t)\rangle$는 시간에만 의존하니까

$$i\hbar \frac{d}{dt} |\psi(t)\rangle = H |\psi(t)\rangle \tag{3-2-4}$$

가 돼.

파인먼은 어떤 특정한 시각 t_0에서 전자의 양자 상태가 $|\psi(t_0)\rangle$으로 주어졌을 때 시각 t에서의 양자 상태 $|\psi(t)\rangle$는

$$|\psi(t)\rangle = U(t) |\psi(t_0)\rangle \tag{3-2-5}$$

으로 주어진다고 보았어.

물리군 $U(t)$는 뭔가요?

정교수 진화 연산자라고 불러.

물리군 시각 t_0일 때의 양자 상태가 시각 t일 때의 양자 상태로 진화하는 과정을 묘사하는 연산자군요.

정교수 그렇지. 자세히 설명해 볼게.

식 (3-2-5)를 식 (3-2-4)에 넣으면

$$i\hbar\frac{d}{dt}(U(t)\mid\psi(t_0)\rangle) = H(U(t)\mid\psi(t_0)\rangle)$$

또는

$$i\hbar\frac{d}{dt}U = HU \tag{3-2-6}$$

이다. 이 식을 풀면

$$U(t) = Ce^{-\frac{i}{\hbar}tH}$$

이 된다. 그러니까

$$\mid\psi(t)\rangle = Ce^{-\frac{i}{\hbar}tH}\mid\psi(t_0)\rangle$$

이다. 이 식의 양변에 $t = t_0$을 대입하면

$$\mid\psi(t_0)\rangle = Ce^{-\frac{i}{\hbar}t_0 H}\mid\psi(t_0)\rangle$$

이므로

$$C = e^{\frac{i}{\hbar}t_0 H}$$

이 된다. 즉,

$$\mid\psi(t)\rangle = e^{-\frac{i}{\hbar}(t-t_0)H}\mid\psi(t_0)\rangle \tag{3-2-7}$$

으로 쓸 수 있다. 따라서 진화 연산자는

$$U(t) = e^{-\frac{i}{\hbar}(t-t_0)H} \tag{3-2-8}$$

이고, 식 (3-2-7)에 대응하는 브라벡터는

$$\langle \psi(t) | = \langle \psi(t_0) | e^{\frac{i}{\hbar}(t-t_0)H} \tag{3-2-9}$$

이 된다.

이제 시간에 의존하지 않는 파동함수를 시각이 $t = 0$일 때의 파동함수라고 하면

$$\psi(x) = \psi(x, 0) = \langle x | \psi \rangle$$

이다. 여기서

$$| \psi \rangle = | \psi(0) \rangle$$

을 의미한다. 이때

$$\psi(x, t) = \langle x | \psi(t) \rangle = \langle x | e^{-\frac{i}{\hbar}tH} | \psi(0) \rangle = \langle x | e^{-\frac{i}{\hbar}tH} | \psi \rangle$$

이고,

$$\psi(x, t) = \langle x, t | \psi \rangle$$

이므로

$$\langle x, t | = \langle x | e^{-\frac{i}{\hbar}tH} \qquad (3-2-10)$$

임을 알 수 있다. 이 식에 수반을 취하면

$$| x, t \rangle = e^{\frac{i}{\hbar}tH} | x \rangle \qquad (3-2-11)$$

가 된다. 여기서 $| x, t \rangle$는 시각 t일 때 전자의 위치 상태를 나타낸다. 그러므로 주어진 시각 t일 때

$$\int | x, t \rangle \langle x, t | dx = I \qquad (3-2-12)$$

가 성립한다.

하위헌스의 원리 _ 과거의 정보로부터 현재가 결정되다

정교수   파인먼의 경로 적분을 이해하려면 먼저 하위헌스의 원리를 알아야 한다네.
물리군   그건 뭐죠?
정교수   우선 물리학자 하위헌스를 소개할게.

하위헌스(Christiaan Huygens, 1629~1695)

하위헌스는 1629년 4월 14일 네덜란드 헤이그의 부유하고 영향력 있는 가정에서 콘스탄테인 하위헌스(Constantijn Huygens)의 둘째 아들로 태어났다. 콘스탄테인 하위헌스는 시인이자 음악가일 뿐만 아니라 오라녜 가문의 외교관이자 고문이었다. 그는 유럽 전역의 지식인들과 서신을 주고받았다. 그의 친구로는 갈릴레오 갈릴레이, 마랭 메르센, 르네 데카르트 등이 있다.

하위헌스의 가족. 오른쪽 위가 하위헌스, 가운데가 아버지

어릴 적부터 방앗간과 다른 기계들의 미니어처를 가지고 노는 것을 좋아했던 하위헌스는 16세가 될 때까지 집에서 교육받았다. 그는 아버지로부터 언어, 음악, 역사, 지리학, 수학, 논리학, 수사학을 배우고 춤, 펜싱, 승마를 하는 교양 교육을 받았다.

16세가 되던 해, 콘스탄테인은 하위헌스를 레이던 대학에 보내 법학과 수학을 배우게 했다. 하위헌스는 1645년 5월부터 1647년 3월까지 그곳에서 공부했다. 이후에 그는 아버지가 큐레이터로 있던 브레다에 새로 설립된 오라녜 칼리지에서 공부를 계속했다. 대학에 다니는 동안에는 법학자 요한 헨리크 다우버르(Johann Henryk Dauber)의 집에 살았고, 영어 강사 존 펠(John Pell)과 함께 수학 수업을 들었다.

1649년 8월에 학업을 마친 하위헌스는 브레다를 떠나 나사우 공작 헨리와 함께 외교관으로 일했다. 아버지 콘스탄테인은 아들이 외교관이 되기를 바랐지만, 하위헌스는 외교관에는 관심이 없었다.

1646년 당시 레이던 대학 학생이었던 하위헌스는 아버지의 친구인 마랭 메르센과 수학에 관한 서신을 주고받기 시작했다. 메르센은 하위헌스를 아르키메데스에 비유했다. 다음 2년(1647~1648) 동안 하위헌스는 자유낙하 법칙에 대한 수학적 증명, 원 구적법 주장, 타원 및 사이클로이드를 연구했다. 그는 훗날 1690년에 현수선에 대한 아이디어를 냈다.

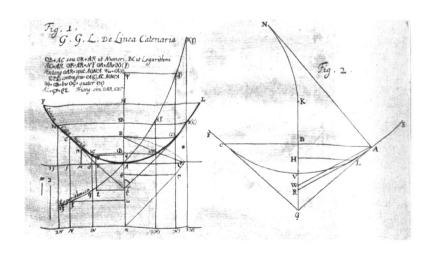

1654년 헤이그에 있는 아버지의 집으로 돌아온 뒤에야 하위헌스는 연구에 전념할 수 있었다. 동시대 사람들과 마찬가지로 그 또한 종종 자신의 연구 결과와 발견을 발표하는 데 시간이 오래 걸렸다. 대신 편지로 지인들에게 그 내용을 알렸다.

1651년에서 1657년 사이에 하위헌스는 해석기하학에 대한 연구 결과를 발표해 수학자들 사이에서 유명해졌다. 같은 시기에 그는 데카르트의 충돌 법칙에 의문을 제기하면서 올바른 충돌 문제를 다루었고, 그 결과는 운동량 보존법칙으로 불렸다.

하위헌스는 1655년에 타이탄을 토성의 위성 중 하나로 처음 식별했다. 1657년에는 진자시계를 발명했으며, 1659년에 토성의 이상한

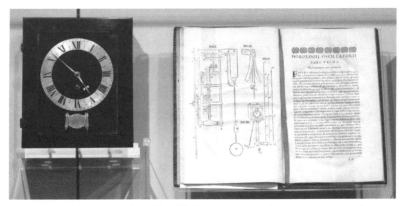

하위헌스가 발명한 진자시계와 그의 논문(출처: Rob Koopman/Wikimedia Commons)

모습을 고리 때문이라고 설명했다. 이 모든 발견은 그에게 유럽 전역에서 명성을 가져다주었다.

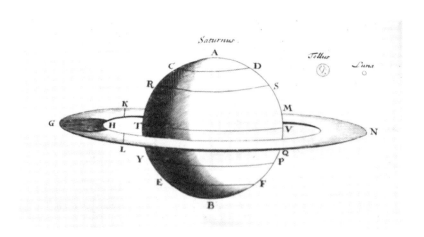

1666년 하위헌스는 프랑스 루이 14세로부터 새로운 프랑스 과학 아카데미의 지도자 직책을 맡아달라는 초청을 받고 파리로 이주했다. 그러나 심각한 우울증을 앓은 후 1681년에 헤이그로 돌아왔다.

1689년 6월 12일, 하위헌스는 세 번째 영국 방문에서 아이작 뉴턴을 직접 만났다. 두 사람은 저항이 있을 때의 물체 운동에 대해 논의했고, 하위헌스가 집으로 돌아간 후에도 서신을 통해 논의가 계속되었다.

물리군 하위헌스의 원리를 설명해 주세요.
정교수 하위헌스의 원리는 파동의 전파에 대한 원리야. 과거의 파면으로부터 현재의 파면이 만들어지지.

물리군　어떻게요?

정교수　과거의 파면 위 모든 점은 새로운 점파원이 되고, 이 점파원에서 만들어진 구면파들의 파면에 대한 공통접선이 새로운 파면이 돼. 다음 그림을 보면 알 수 있어.

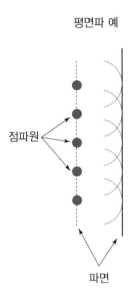

평면파 예

물리군　그러니까 과거에 대한 정보로부터 현재가 결정되는 방식으로 파동이 전파되는군요.

정교수　그렇다네.

경로 적분 _ 전자를 발견할 확률 구하기

정교수 파인먼이 생각한 경로 적분은 하위헌스의 원리처럼 과거의 정보가 모여서 현재의 정보를 만들어내는 과정과 비슷해. 그러니까 앞에서 공부한 시간-위치 켓벡터 $|x, t\rangle$를 사용해야 하지. 자네가 시각 t_f일 때 전자를 x_f에서 관측했다고 가정해 보게. 그러면 관측한 전자의 시간-위치 켓벡터는

$$|x_f, t_f\rangle$$

가 돼. 전자는 공간 속에서 시간에 따라 이동하지?

물리군 그렇죠.

정교수 따라서 시간에 따라 위치가 달라질 거야. 자네가 관측한 시각보다 과거의 어떤 시각 t_i를 생각해 봐. 이 시각에서 전자의 위치를 x_i로 두게. 그런데 자네는 현재 시각인 t_f에서 전자의 위치는 관측했지만 이 전자가 과거 시각 t_i에는 어디에 있는지 모르잖아? 그러니까 x_i는 모든 실수가 될 수 있어.

물리군 무한개의 $|x_i, t_i\rangle$를 고려해야 하는군요. 이때

$$-\infty < x_i < \infty$$

가 되고요.

정교수 맞아. 시각 t_i에서의 시간-위치 켓벡터는

$$\int |x_i, t_i\rangle\langle x_i, t_i| dx_i = I \quad (3\text{-}4\text{-}1)$$

$$\langle x_i, t_i | x'_i, t_i \rangle = \delta(x_i - x'_i) \quad (3\text{-}4\text{-}2)$$

를 만족하지. 파인먼은 현재 시각에서의 시간-위치 켓벡터가 과거 시각에서의 시간-위치 켓벡터의 중첩으로 이루어진다고 보았어. 즉,

$$|x_f, t_f\rangle = \int a_{x_i} |x_i, t_i\rangle dx_i \quad (3\text{-}4\text{-}3)$$

가 되지. 식 (3-4-1)을 이용하면

$$a_{x_i} = \langle x_i, t_i | x_f, t_f \rangle \quad (3\text{-}4\text{-}4)$$

이니까 식 (3-4-3)은

$$|x_f, t_f\rangle = \int \langle x_i, t_i | x_f, t_f\rangle |x_i, t_i\rangle dx_i \quad (3\text{-}4\text{-}5)$$

로 쓸 수 있어.

물리군 a_{x_i}의 의미는 뭐죠?

정교수 조금 복잡하지? 더 간단한 경우를 먼저 고려해 볼게. 시각 t_i일 때 가능한 위치가 x_1, x_2의 두 가지라고 하면

$$x_i = x_1, x_2$$

라네. 이때는 가능한 위치가 두 가지이므로 적분이 아니라 덧셈에 의

해 중첩돼. 그러니까

$$|x_f, t_f\rangle = a_{x_1}|x_1, t_i\rangle + a_{x_2}|x_2, t_i\rangle$$

이지. 즉, 두 상태의 중첩이야. 이때

$$a_{x_1} = \langle x_1, t_i | x_f, t_f \rangle$$

이고,

$$|a_{x_1}|^2 = |\langle x_1, t_i | x_f, t_f \rangle|^2$$

$$= (t_i \text{일 때 } x_1 \text{에 있던 전자가 } t_f \text{일 때 } x_f \text{에 있을 확률})$$

이 돼. 마찬가지로

$$a_{x_2} = \langle x_2, t_i | x_f, t_f \rangle$$

이고,

$$|a_{x_2}|^2 = |\langle x_2, t_i | x_f, t_f \rangle|^2$$

$$= (t_i \text{일 때 } x_2 \text{에 있던 전자가 } t_f \text{일 때 } x_f \text{에 있을 확률})$$

이지.

물리군 전자가 이사 갈 확률이군요.

정교수 그렇지. 그래서 이 확률을 전이확률이라고 불러. 식 (3-4-4) 로부터 시간-위치 켓벡터 $|x_i, t_i\rangle$에서 $|x_f, t_f\rangle$로의 전이확률은

$$|a_{x_i}|^2 = |\langle x_i, t_i | x_f, t_f \rangle|^2 = |\langle x_f, t_f | x_i, t_i \rangle|^2$$

이 돼.

물리군 이것을 파동함수로 나타낼 수 있나요?

정교수 물론이야. 임의의 상태 켓벡터를 $|\psi\rangle$라고 할 때, 식 (3-4-5)의 수반을 취하면

$$\langle x_f, t_f | = \int \langle x_f, t_f | x_i, t_i \rangle \langle x_i, t_i | dx_i \tag{3-4-6}$$

이지. 양변의 오른쪽에 $|\psi\rangle$를 곱하면

$$\langle x_f, t_f | \psi \rangle = \int \langle x_f, t_f | x_i, t_i \rangle \langle x_i, t_i | \psi \rangle dx_i$$

이므로

$$\psi(x_f, t_f) = \int \langle x_f, t_f | x_i, t_i \rangle \psi(x_i, t_i) dx_i \tag{3-4-7}$$

가 돼. 그러니까 $\psi(x_i, t_i)$는 시각 t_i일 때 위치 x_i에 있는 전자를 묘사하는 파동함수이고, $\psi(x_f, t_f)$는 시각 t_f일 때 위치 x_f에 있는 전자를 묘사하는 파동함수야. 이때

$$K(x_f, t_f | x_i, t_i) = \langle x_f, t_f | x_i, t_i \rangle$$

로 쓰고 전파인자라고 불러. 따라서 식 (3-4-7)은

$$\psi(x_f, t_f) = \int K(x_f, t_f \mid x_i, t_i) \psi(x_i, t_i) \, dx_i \qquad (3\text{-}4\text{-}8)$$

이지. 그리고 시각 t_i일 때 위치 x_i에서 전자를 발견할 확률은

$|\psi(x_i, t_i)|^2$

이고, 시각 t_f일 때 위치 x_f에서 전자를 발견할 확률은

$|\psi(x_f, t_f)|^2$

이야.

물리군 전파인자를 구하면 확률이 어떻게 변하는지 적분을 통해 알 수 있군요.

정교수 맞아. 과거 시각에서 모든 가능한 위치의 파동함수를 섞어서 (중첩시켜서) 현재 시각일 때 어떤 위치의 파동함수를 구할 수 있지. 이때 식 (3-4-8)을 경로 적분이라고 불러. 경로 적분을 통해 파동함수를 구할 수 있으면 전자를 발견할 확률도 구할 수 있지.

물리군 뉴턴 역학에서는 경로 적분이 필요 없잖아요?

정교수 물론이야. 하지만 양자역학에서는 어느 장소를 통해 최종 장소로 도달했는지 알 수 없어. 모든 가능한 경우를 헤아리면 올바른 결론에 도달할 수 있지. 그래서 경로 적분을 하는 거야.

이제 t_i와 t_f 사이의 시각 t를 생각하자. 시각 t일 때 가능한 시간-위치 켓벡터를 $|x, t\rangle$라고 하면

$$\int |x,t\rangle\langle x,t|\,dx = I$$

이다. 이 식을 이용하면

$$\psi(x_f, t_f) = \int \langle x_f, t_f | dx_i \int |x,t\rangle\langle x,t|x_i,t_i\rangle\langle x_i,t_i|\psi\rangle\,dx$$

$$= \int dx_i \int \langle x_f, t_f|x,t\rangle\langle x,t|x_i,t_i\rangle\langle x_i,t_i|\psi\rangle\,dx$$

$$= \int \left[\int \langle x_f, t_f|x,t\rangle\langle x,t|x_i,t_i\rangle\,dx\right]\langle x_i,t_i|\psi\rangle\,dx_i$$

가 된다. 그러므로 전파인자에 대한 다음 성질을 얻을 수 있다.

$$K(x_f, t_f | x_i, t_i) = \int K(x_f, t_f | x, t) K(x, t | x_i, t_i)\,dx \qquad (3\text{-}4\text{-}9)$$

물리군 식 (3-4-9)는 이해하겠는데 감이 잘 안 와요.

정교수 다음과 같은 이중 슬릿 실험을 생각해 보세.

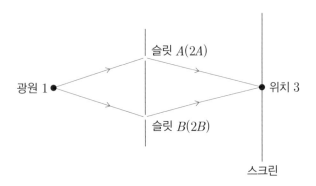

광원에서 광자가 나오는 순간의 시각을 t_i, 광자가 슬릿 A 또는 슬릿 B를 통과하는 시각을 t, 광자가 스크린의 위치 x에 도달하는 시각을 t_f라고 하자. 시각 t_i일 때 허용된 위치는 1 하나뿐이고, 시각 t_f일 때 허용된 위치는 3 하나뿐이다. 그런데 시각 t일 때 허용 가능한 위치는 $2A$ 또는 $2B$이다. 그러니까

$$|2A,t\rangle\langle 2A,t|+|2B,t\rangle\langle 2B,t|=I \tag{3-4-10}$$

이다. 따라서 3에서의 파동함수를 ψ_3이라고 하면

$$\psi_3=\langle 3,t_f|\psi\rangle \tag{3-4-11}$$

가 된다. 식 (3-4-10)을 식 (3-4-11)에 넣으면

$$\psi_3=\langle 3,t_f|(|2A,t\rangle\langle 2A,t|+|2B,t\rangle\langle 2B,t|)\psi\rangle$$

$$=\langle 3,t_f|2A,t\rangle\langle 2A,t|\psi\rangle+\langle 3,t_f|2B,t\rangle\langle 2B,t|\psi\rangle$$

이다. $2A$, $2B$에서의 파동함수를 각각 ψ_{2A}, ψ_{2B}라고 하면

$$\psi_{2A}=\langle 2A,t|\psi\rangle$$

$$\psi_{2B}=\langle 2B,t|\psi\rangle$$

로 쓸 수 있다. 즉,

$$\psi_3=K(3,t_f|2A,t)\psi_{2A}+K(3,t_f|2B,t)\psi_{2B} \tag{3-4-12}$$

이다. 그리고 1에서 2A로 가는 경로는 하나뿐이므로

$$\psi_{2A} = \langle 2A, t | \psi \rangle$$

$$= \langle 2A, t | 1, t_i \rangle \langle 1, t_i | \psi \rangle$$

$$= K(2A, t | 1, t_i) \psi_1$$

이 된다. 여기서 ψ_1은 1에서의 파동함수이다. 마찬가지로 다음과 같다.

$$\psi_{2B} = \langle 2B, t | \psi \rangle$$

$$= \langle 2B, t | 1, t_i \rangle \langle 1, t_i | \psi \rangle$$

$$= K(2B, t | 1, t_i) \psi_1$$

따라서 식 (3-4-12)는

$$\psi_3 = [K(3, t_f | 2A, t) K(2A, t | 1, t_i) + K(3, t_f | 2B, t) K(2B, t | 1, t_i)] \psi_1$$
(3-4-13)

이라고 쓸 수 있다.

이번에는 위치 1과 위치 3을 비교해 보자. 그러면

$$\psi_3 = \langle 3, t_f | \psi \rangle$$

$$= \langle 3, t_f | 1, t_i \rangle \langle 1, t_i | \psi \rangle$$

$$= K(3, t_f | 1, t_i) \psi_1 \tag{3-4-14}$$

이 된다. 식 (3-4-13)과 (3-4-14)를 비교하면

$$K(3, t_f \mid 1, t_i)$$

$$= K(3, t_f \mid 2A, t)K(2A, t \mid 1, t_i) + K(3, t_f \mid 2B, t)K(2B, t \mid 1, t_i)$$

이다. 그러니까 확률은

$$|K(3, t_f \mid 1, t_i)|^2$$

$$\neq |K(3, t_f \mid 2A, t)K(2A, t \mid 1, t_i)|^2 + |K(3, t_f \mid 2B, t)K(2B, t \mid 1, t_i)|^2$$

을 만족한다. 이렇게 부등식이 성립하는 이유는 바로 간섭 때문이다. 즉, 슬릿 A를 통과한 빛과 슬릿 B를 통과한 빛이 간섭을 일으키기 때문에 각각의 슬릿을 통과하는 확률의 합이 3에서의 확률과 같지 않은 것이다.

물리군   그렇군요.

경로 적분을 구하는 방법 _ 모든 가능한 경로에 대해

정교수   이제 경로 적분을 일반적으로 구하는 방법을 소개하려고 해. 수식이 조금 복잡할 거야.

물리군   집중할게요.

정교수   시각 t_i에서의 전자가 시각 t_f로 전이되는 경우를 생각할게. 시각 t_i일 때 전자의 위치는 x_i, 시각 t_f일 때 전자의 위치는 x_f라고 할 때, 두 시각 사이의 시간은

$$t_f - t_i$$

로 쓸 수 있어. 이 시간을 일정한 간격 τ로 4등분하면

$$\tau = \frac{t_f - t_i}{4}$$

이지. 이때 등분으로 생긴 시각을 각각 t_1, t_2, t_3이라고 하세.

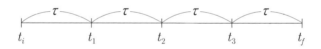

그리고 t_1, t_2, t_3일 때 전자의 위치를 각각 x_1, x_2, x_3으로 놓을 거야.

물리군 x_1, x_2, x_3은 알 수 있나요?

정교수   아니. t_1, t_2, t_3일 때 전자가 어디에 있는지는 알 수 없어. 처음 위치와 나중 위치만 알 뿐이지.

물리군   전자가 어디 있는지 모르는데 어떻게 경로 적분을 하죠?

정교수   모든 가능한 경로에 대해 경로 적분을 하는 거야.

다음 그림은 무한히 많은 경로 중 하나를 나타낸 것이다.

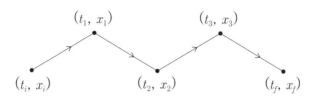

이때

$$\int |x_1, t_1\rangle\langle x_1, t_1| dx_1 = I$$

$$\int |x_2, t_2\rangle\langle x_2, t_2| dx_2 = I$$

$$\int |x_3, t_3\rangle\langle x_3, t_3| dx_3 = I$$

가 성립한다. 그러니까

$$\langle x_f, t_f | x_i, t_i \rangle$$

$$= \langle x_f, t_f | I \times I \times I | x_i, t_i \rangle$$

$$= \langle x_f, t_f | \left(\int |x_3, t_3\rangle\langle x_3, t_3| dx_3\right)\left(\int |x_2, t_2\rangle\langle x_2, t_2| dx_2\right)\left(\int |x_1, t_1\rangle\langle x_1, t_1| dx_1\right) | x_i, t_i \rangle$$

$$= \iiint \langle x_f, t_f | x_3, t_3\rangle\langle x_3, t_3 | x_2, t_2\rangle\langle x_2, t_2 | x_1, t_1\rangle\langle x_1, t_1 | x_i, t_i\rangle dx_1 dx_2 dx_3$$

이 된다.

이번에는 시간을 일정한 간격 τ로 $(n+1)$등분하자. 그러면

$$\tau = \frac{t_f - t_i}{n+1}$$

이다. 그러므로 등분의 개수를 엄청나게 크게 택하면 τ는 거의 0에 가까운 작은 값이 된다. 이때 등분으로 생긴 시각을 각각

$t_1, t_2, \cdots, t_n$

이라고 하자.

시각 t_j일 때 전자의 위치를 x_j라고 하자. 이 경우 $j = 1, 2, 3, \cdots, n$이다. 여기서

$$\int |x_1, t_1\rangle\langle x_1, t_1| \, dx_1 = I$$

$$\int |x_2, t_2\rangle\langle x_2, t_2| \, dx_2 = I$$

$$\int |x_3, t_3\rangle\langle x_3, t_3| \, dx_3 = I$$

$$\vdots$$

$$\int |x_n, t_n\rangle\langle x_n, t_n| \, dx_n = I$$

를 이용하면

$$\langle x_f, t_f | x_i, t_i \rangle$$

$$= \int \cdots \int \langle x_f, t_f | x_n, t_n \rangle \langle x_n, t_n | x_{n-1}, t_{n-1} \rangle \cdots \langle x_1, t_1 | x_i, t_i \rangle dx_1 dx_2 \cdots dx_n$$

(3-5-1)

이 된다.

이제 다음과 같이 놓자.

$$x_f = x_{n+1}, \qquad t_f = t_{n+1}$$

$$x_i = x_0, \qquad t_i = t_0$$

이때

$$\langle x_f, t_f | x_i, t_i \rangle$$

$$= \int \cdots \int \langle x_{n+1}, t_{n+1} | x_n, t_n \rangle \langle x_n, t_n | x_{n-1}, t_{n-1} \rangle \cdots \langle x_1, t_1 | x_0, t_0 \rangle dx_1 dx_2 \cdots dx_n$$

(3-5-2)

으로 쓸 수 있다. 여기서 다음과 같이 조금 간단한 기호를 사용하자.

$$|x_j, t_j \rangle = |j\rangle$$

$$\langle x_j, t_j| = \langle j|$$

$$K(x_f, t_f | x_i, t_i) = K(f|i) = \langle x_f, t_f | x_i, t_i \rangle = \langle f|i \rangle$$

그리고 적분 기호를 여러 번 쓰는 대신 한 번만 쓰고 여러 번 적분하는 것으로 약속하자. 또한

$$dx_1 dx_2 \cdots dx_n = Dx(1 \to n)$$

으로 놓자. 그러면

$$\langle f | i \rangle = \langle n+1 | 0 \rangle = \int Dx(1 \to n) \langle n+1 | n \rangle \langle n | n-1 \rangle \cdots \langle 1 | 0 \rangle$$

(3-5-3)

이다. 진화 연산자를 이용하면

$$\langle j+1 | j \rangle = \langle x_{j+1} | e^{-\frac{i}{\hbar}Ht_{j+1}} e^{\frac{i}{\hbar}Ht_j} | x_j \rangle$$

$$= \langle x_{j+1} | e^{-\frac{i}{\hbar}H(t_{j+1}-t_j)} | x_j \rangle$$

$$= \langle x_{j+1} | e^{-\frac{i}{\hbar}H\tau} | x_j \rangle \qquad (3\text{-}5\text{-}4)$$

가 된다. τ가 아주 작기 때문에

$$\langle x_{j+1} | e^{-\frac{i}{\hbar}H\tau} | x_j \rangle \approx \langle x_{j+1} | 1 - \frac{i}{\hbar}H\tau | x_j \rangle$$

$$= \langle x_{j+1} | x_j \rangle + \langle x_{j+1} | -\frac{i}{\hbar}H\tau | x_j \rangle$$

$$= \delta(x_{j+1} - x_j) - \frac{i}{\hbar} \langle x_{j+1} | H\tau | x_j \rangle$$

$$= \frac{1}{2\pi\hbar} \int e^{\frac{i}{\hbar}p_j(x_{j+1}-x_j)} dp_j - \frac{i}{\hbar}\tau \langle x_{j+1} | \hat{H} | x_j \rangle$$

(3-5-5)

이고,

$$H = \frac{1}{2m}\hat{p}^2 + V(\hat{x})$$

이므로

$$\langle x_{j+1} | \hat{H} | x_j \rangle = \langle x_{j+1} | \frac{1}{2m}\hat{p}^2 | x_j \rangle + \langle x_{j+1} | V(\hat{x}) | x_j \rangle \quad (3\text{-}5\text{-}6)$$

가 된다. 여기서

$$\langle x_{j+1} | \frac{1}{2m}\hat{p}^2 | x_j \rangle$$

$$= \frac{1}{2m}\int \langle x_{j+1} | p' \rangle \langle p' | \hat{p}^2 | p \rangle \langle p | x_j \rangle dp'dp$$

$$= \frac{1}{2m}\int \langle x_{j+1} | p' \rangle \langle p' | p^2 | p \rangle \langle p | x_j \rangle dp'dp$$

$$= \frac{1}{2m}\int p^2 \langle x_{j+1} | p' \rangle \langle p' | p \rangle \langle p | x_j \rangle dp'dp$$

$$= \frac{1}{2m}\int p^2 \langle x_{j+1} | p' \rangle \delta(p'-p) \langle p | x_j \rangle dp'dp$$

$$= \frac{1}{2m}\int p^2 \langle x_{j+1} | p \rangle \langle p | x_j \rangle dp$$

$$= \frac{1}{2m}\int p_j^2 \langle x_{j+1} | p_j \rangle \langle p_j | x_j \rangle dp_j$$

$$= \int \frac{1}{h} e^{\frac{i}{\hbar}p_j(x_{j+1}-x_j)} \frac{p_j^2}{2m} dp_j \quad (3\text{-}5\text{-}7)$$

이고,

$$\langle x_{j+1} | V(\hat{x}) | x_j \rangle = V(x_j) \langle x_{j+1} | x_j \rangle$$

$$= \int \frac{1}{h} e^{\frac{i}{\hbar} p_j (x_{j+1} - x_j)} V(x_j) dp_j \qquad (3-5-8)$$

이다. 식 (3-5-7)과 (3-5-8)을 식 (3-5-6)에 넣으면 다음과 같다.

$$\langle x_{j+1} | \hat{H} | x_j \rangle = \int \frac{1}{h} e^{\frac{i}{\hbar} p_j (x_{j+1} - x_j)} H(p_j, x_j) dp_j \qquad (3-5-9)$$

이때

$$H(p_j, x_j) = \frac{p_j^2}{2m} + V(x_j) \qquad (3-5-10)$$

이다. 따라서

$$\langle j+1 | j \rangle \approx \frac{1}{2\pi\hbar} \int e^{\frac{i}{\hbar} p_j (x_{j+1} - x_j)} dp_j - \frac{i}{\hbar} \tau \int \frac{1}{h} e^{\frac{i}{\hbar} p_j (x_{j+1} - x_j)} H(p_j, x_j) dp_j$$

$$= \frac{1}{h} \int e^{\frac{i}{\hbar} p_j (x_{j+1} - x_j)} dp_j - \frac{i}{\hbar} \tau \int \frac{1}{h} e^{\frac{i}{\hbar} p_j (x_{j+1} - x_j)} H(p_j, x_j) dp_j$$

$$= \frac{1}{h} \int e^{\frac{i}{\hbar} p_j (x_{j+1} - x_j)} \left(1 - \frac{i}{\hbar} \tau H(p_j, x_j) \right) dp_j \qquad (3-5-11)$$

가 된다. τ가 아주 작기 때문에 다음과 같이 쓸 수 있다.

$$\langle j+1 | j \rangle = \frac{1}{h} \int e^{\frac{i}{\hbar} p_j(x_{j+1}-x_j)} e^{-\frac{i}{\hbar}\tau H(p_j,x_j)} dp_j$$

$$= \frac{1}{h} \int e^{\frac{i}{\hbar}[p_j(x_{j+1}-x_j)-\tau H(p_j,x_j)]} dp_j \quad (3\text{-}5\text{-}12)$$

식 (3-5-12)를 식 (3-5-3)에 넣으면

$$K(f|i) = \langle f | i \rangle = \int Dx(1 \to n) Dp(0 \to n) e^{\frac{i}{\hbar}\sum_{j=0}^{n}[p_j(x_{j+1}-x_j)-\tau H(p_j,x_j)]}$$
$$(3\text{-}5\text{-}13)$$

이다. 여기서 다음을 알 수 있다.

$$Dp(0 \to n) = \frac{1}{h^{n+1}} dp_0 dp_1 dp_2 \cdots dp_n \quad (3\text{-}5\text{-}14)$$

전파인자의 계산 _ 전자가 힘을 받지 않는 경우와 힘을 받는 경우

정교수 전파인자를 계산하기 위해 식 (3-5-13)에서

$$I_n = \int Dp(0 \to n) e^{\frac{i}{\hbar}\sum_{j=0}^{n}\left[p_j(x_{j+1}-x_j)-\tau\frac{p_j^2}{2m}-\tau V(x_j)\right]} \quad (3\text{-}6\text{-}1)$$

으로 놓을 거야. 이 적분을 계산하려면 다음 공식을 알아야 해.

$$\int_{-\infty}^{\infty} e^{-ax^2} dx = \sqrt{\frac{\pi}{a}} \quad (a > 0) \quad (3\text{-}6\text{-}2)$$

물리군 어떻게 나온 공식이죠?

정교수 이 적분을

$$I = \int_{-\infty}^{\infty} e^{-ax^2} dx$$

라고 두면

$$I^2 = \left(\int_{-\infty}^{\infty} e^{-ax^2} dx \right) \left(\int_{-\infty}^{\infty} e^{-ay^2} dy \right)$$

$$= \int_{-\infty}^{\infty} \int_{-\infty}^{\infty} e^{-a(x^2+y^2)} dxdy$$

가 돼. 이제 극좌표를 사용해 볼까? 극좌표계는

$$x = r\cos\theta$$

$$y = r\sin\theta$$

로 나타낼 수 있어. 여기서

$$r = \sqrt{x^2 + y^2}$$

으로 원점으로부터의 거리이고 θ는 0부터 2π까지를 나타내는 각변수야. 그런데 데카르트 좌표계에서의 넓이 요소 $dxdy$를 극좌표로 나타내면 $rdrd\theta$가 되거든. 그러니까 다음과 같아.

$$x^2 + y^2 = r^2$$

$$dxdy = rdrd\theta$$

r은 0에서 ∞까지 변하고 θ는 0에서 2π까지 변하니까

$$I^2 = \int_{r=0}^{\infty} \int_{\theta=0}^{2\pi} re^{-ar^2} drd\theta$$

$$= \int_{r=0}^{\infty} re^{-ar^2} dr \int_{\theta=0}^{2\pi} d\theta$$

$$= \frac{\pi}{a}$$

로부터

$$\int_{-\infty}^{\infty} e^{-ax^2} dx = \sqrt{\frac{\pi}{a}}$$

가 되지. 이 적분 공식을 이용하면 다음 공식을 만들 수 있어.

$$\int_{-\infty}^{\infty} e^{-ax^2+bx+c} dx = e^{\frac{b^2}{4a}+c} \sqrt{\frac{\pi}{a}} \qquad (3\text{-}6\text{-}3)$$

물리군 이건 어떻게 증명하나요?

정교수 방법은 간단해.

$$-ax^2 + bx + c = -a\left(x - \frac{b}{2a}\right)^2 + \frac{b^2}{4a} + c$$

이니까

$$\int_{-\infty}^{\infty} e^{-ax^2+bx+c} dx = \int_{-\infty}^{\infty} e^{-a\left(x-\frac{b}{2a}\right)^2 + \frac{b^2}{4a}+c} dx$$

$$= e^{\frac{b^2}{4a}+c} \int_{-\infty}^{\infty} e^{-a\left(x-\frac{b}{2a}\right)^2} dx$$

이지. 적분에서

$$x - \frac{b}{2a} = y$$

로 치환하면

$$\int_{-\infty}^{\infty} e^{-ax^2+bx+c} dx = e^{\frac{b^2}{4a}+c} \int_{-\infty}^{\infty} e^{-ay^2} dy$$

$$= e^{\frac{b^2}{4a}+c} \sqrt{\frac{\pi}{a}}$$

가 돼.

물리군 그렇군요.

정교수 이제 식 (3-6-1)에 $n=0$을 넣으면

$$I_0 = \int \frac{1}{h} e^{\frac{i}{\hbar}\left[p_0(x_1-x_0) - \tau \frac{p_0^2}{2m} - \tau V(x_0)\right]} dp_0$$

$$= \int \frac{1}{h} e^{-\frac{i}{\hbar}\tau \frac{p_0^2}{2m} + \frac{i}{\hbar}p_0(x_1-x_0) - \frac{i}{\hbar}\tau V(x_0)} dp_0$$

이지? 이것은 식 (3-6-3)에서

$$a = \frac{i\tau}{2m\hbar}, \quad b = \frac{i}{\hbar}(x_1 - x_0), \quad c = -\frac{i}{\hbar}\tau V(x_0)$$

인 경우야. 그러니까

$$I_0 = \sqrt{\frac{m}{ih\tau}} e^{\frac{im}{2\hbar\tau}(x_1-x_0)^2 - \frac{i}{\hbar}\tau V(x_0)}$$

이 돼. 같은 방법으로 계산하면

$$I_n = \left(\sqrt{\frac{m}{ih\tau}}\right)^{n+1} e^{\frac{im}{2\hbar\tau}\sum_{j=0}^{n}(x_{j+1}-x_j)^2 - \frac{i}{\hbar}\tau\sum_{j=0}^{n}V(x_j)}$$

을 구할 수 있어. 따라서

$$K(f\mid i) = \int Dx(1 \to n) \left(\sqrt{\frac{m}{ih\tau}}\right)^{n+1} e^{\frac{im}{2\hbar\tau}\sum_{j=0}^{n}(x_{j+1}-x_j)^2 - \frac{i}{\hbar}\tau\sum_{j=0}^{n}V(x_j)}$$

(3-6-4)

이 되지.

먼저 전자가 힘을 받지 않는 경우의 전파인자를 구해 보자. 전자가 힘을 받지 않으면 전자의 운동량이 달라지지 않고 등속직선운동을 한다. 즉, 힘을 받지 않는 경우는 퍼텐셜에너지가 0임을 의미한다. 이때 처음 시공간 점 i에서 나중 시공간 점 f까지의 전파인자를 자유전파인자라 부르고 $K_0(f\mid i)$로 쓰는데

$$K_0(f\mid i) = \int Dx(1 \to n) \left(\sqrt{\frac{m}{ih\tau}}\right)^{n+1} e^{\frac{im}{2\hbar\tau}\sum_{j=0}^{n}(x_{j+1}-x_j)^2}$$

이다. 이것을 다음과 같은 파인먼 다이어그램으로 나타낸다.

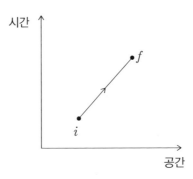

이제 자유전파인자를 구해 보자.

$$J_n = \int Dx\,(1 \to n) \left(\sqrt{\frac{m}{ih\tau}}\right)^{n+1} e^{\frac{im}{2\hbar\tau}\sum_{j=0}^{n}(x_{j+1}-x_j)^2}$$

으로 두고, 먼저 $n = 1$인 경우를 보자. 이때

$$J_1 = \int_{-\infty}^{\infty} \left(\sqrt{\frac{m}{ih\tau}}\right)^2 e^{\frac{im}{2\hbar\tau}[(x_1-x_0)^2+(x_2-x_1)^2]} dx_1$$

$$= \left(\sqrt{\frac{m}{ih\tau}}\right)^2 \int_{-\infty}^{\infty} e^{-\frac{m}{\hbar\tau i}x_1^2 + \frac{m}{\hbar\tau i}(x_0+x_2)x_1 - \frac{m}{2\hbar\tau i}(x_0^2+x_2^2)} dx_1$$

이고, 식 (3-6-3)을 이용하면

$$J_1 = \sqrt{\frac{m}{2\pi\hbar i(2\tau)}}\, e^{\frac{im(x_2-x_0)^2}{2\hbar(2\tau)}}$$

이 된다. 이 경우

$$2\tau = t_2 - t_0 = t_f - t_i$$

$$x_2 = x_f, \qquad x_0 = x_i$$

이므로

$$J_1 = \sqrt{\frac{m}{2\pi\hbar i(t_f - t_i)}} e^{\frac{im(x_f - x_i)^2}{2\hbar(t_f - t_i)}}$$

이다. 이제 $n = 2$인 경우를 보자.

$$J_2 = \int_{-\infty}^{\infty} dx_1 \int_{-\infty}^{\infty} \left(\sqrt{\frac{m}{i h \tau}}\right)^3 e^{\frac{im}{2\hbar\tau}[(x_1 - x_0)^2 + (x_2 - x_1)^2 + (x_3 - x_2)^2]} dx_2$$

$$= \left(\sqrt{\frac{m}{i h \tau}}\right)^3 \int_{-\infty}^{\infty} e^{\frac{im}{2\hbar\tau}(x_1 - x_0)^2} dx_1 \int_{-\infty}^{\infty} e^{\frac{im}{2\hbar\tau}[(x_2 - x_1)^2 + (x_3 - x_2)^2]} dx_2$$

$$= \left(\sqrt{\frac{m}{i h \tau}}\right)^3 \sqrt{\frac{i h \tau}{2m}} \int_{-\infty}^{\infty} e^{\frac{im}{2\hbar\tau}(x_1 - x_0)^2} e^{\frac{im}{4\hbar\tau}(x_3 - x_1)^2} dx_1$$

$$= \left(\sqrt{\frac{m}{i h \tau}}\right)^3 \sqrt{\frac{i h \tau}{2m}} \int_{-\infty}^{\infty} e^{-\frac{3m}{4\hbar\tau i}x_1^2 + \frac{m}{2\hbar\tau i}(x_3 + 2x_0)x_1 - \frac{m}{4\hbar\tau i}(2x_0^2 + x_3^2)} e^{\frac{im}{4\hbar\tau}(x_3 - x_1)^2} dx_1$$

$$= \left(\sqrt{\frac{m}{i h \tau}}\right)^3 \sqrt{\frac{i h \tau}{2m}} \sqrt{\frac{2 i h \tau}{3m}} e^{\frac{im(x_3 - x_0)^2}{2\hbar(3\tau)}}$$

$$= \sqrt{\frac{m}{2\pi\hbar i(3\tau)}} e^{\frac{im(x_3 - x_0)^2}{2\hbar(3\tau)}}$$

이때

$$3\tau = t_3 - t_0 = t_f - t_i$$

$$x_3 = x_f, \quad x_0 = x_i$$

이므로

$$J_2 = \sqrt{\frac{m}{2\pi\hbar i(t_f - t_i)}} e^{\frac{im(x_f - x_i)^2}{2\hbar(t_f - t_i)}}$$

이다. 일반적으로 임의의 n에 대해

$$J_n = \sqrt{\frac{m}{2\pi\hbar i(t_f - t_i)}} e^{\frac{im(x_f - x_i)^2}{2\hbar(t_f - t_i)}}$$

이므로

$$K_0(f \mid i) = \sqrt{\frac{m}{2\pi\hbar i(t_f - t_i)}} e^{\frac{im(x_f - x_i)^2}{2\hbar(t_f - t_i)}} \quad (3\text{-}6\text{-}5)$$

임을 알 수 있다.

물리군 전자가 힘을 받는 경우는 어떻게 되죠?
정교수 그건 퍼텐셜에너지가 0이 아닌 경우야. 식으로 설명해 볼게.

$$K(f \mid i) = \int Dx(1 \to n) \left(\sqrt{\frac{m}{ih\tau}}\right)^{n+1} e^{\frac{im}{2\hbar\tau}\sum_{j=0}^{n}(x_{j+1}-x_j)^2} e^{-\frac{i}{\hbar}\tau\sum_{j=0}^{n}V(x_j)}$$

$$(3\text{-}6\text{-}6)$$

에서 n을 아주 크게 택하면 τ가 아주 작아져

$$\approx \int Dx(1 \to n)\left(\sqrt{\frac{m}{ih\tau}}\right)^{n+1} e^{\frac{im}{2\hbar\tau}\sum_{j=0}^{n}(x_{j+1}-x_j)^2}\left[1-\frac{i}{\hbar}\tau\sum_{k=0}^{n}V(x_k)\right]$$

$$= K_0(f \mid i) + K_1(f \mid i)$$

가 된다. 여기서

$$K_1(f \mid i) = \int Dx(1 \to n)\left(\sqrt{\frac{m}{ih\tau}}\right)^{n+1} e^{\frac{im}{2\hbar\tau}\sum_{j=0}^{n}(x_{j+1}-x_j)^2}\left[-\frac{i}{\hbar}\tau\sum_{k=0}^{n}V(x_k)\right]$$

$$= -\frac{i}{\hbar}\tau\sum_{k=0}^{n}\int Dx(1 \to n)\left(\sqrt{\frac{m}{ih\tau}}\right)^{n+1} e^{\frac{im}{2\hbar\tau}\sum_{j=0}^{n}(x_{j+1}-x_j)^2}V(x_k)$$

이다. 그런데

$$e^{\frac{im}{2\hbar\tau}\sum_{j=0}^{n}(x_{j+1}-x_j)^2} = e^{\frac{im}{2\hbar\tau}\sum_{j=0}^{k-1}(x_{j+1}-x_j)^2} e^{\frac{im}{2\hbar\tau}\sum_{j=k}^{n}(x_{j+1}-x_j)^2}$$

이므로

$$K_1(f \mid i)$$

$$= -\frac{i}{\hbar}\tau\sum_{k=0}^{n}\int dx_k \int Dx(k \to n)\left(\sqrt{\frac{m}{ih\tau}}\right)^{n-k+1} e^{\frac{im}{2\hbar\tau}\sum_{j=k}^{n}(x_{j+1}-x_j)^2}V(x_k)$$

$$\times \int Dx(1 \to k-1)\left(\sqrt{\frac{m}{ih\tau}}\right)^{k} e^{\frac{im}{2\hbar\tau}\sum_{j=0}^{k-1}(x_{j+1}-x_j)^2}$$

$$= -\frac{i}{\hbar}\tau\sum_{k=0}^{n}\int K_0(f \mid k)V(x_k)K_0(k \mid i)\,dx_k$$

가 된다. 이때 x_k의 위치는 모르기 때문에 모든 가능한 값에 대해 적분을 취한다.

파인먼은 이것을 다음과 같이 파인먼 다이어그램으로 나타냈다.

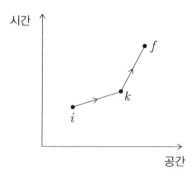

즉, 시각 t_k일 때 위치 x_k에서 힘을 받아 퍼텐셜에너지가 생기고, 이로 인해 운동 방향이 바뀐다. 여기서 $t_i < t_k < t_f$이다.

식 (3-6-6)에서 지수함수를 τ^2항까지 더 전개하면 퍼텐셜에너지의 제곱항은

$$K_2(f|i) = \left(-\frac{i}{\hbar}\tau\right)^2 \sum_{k=0}^{n}\sum_{l=0}^{n} \int dx_k \int K_0(f|l)V(x_l)K_0(l|k)V(x_k)K_0(k|i)\,dx_l$$

과 같다. 예를 들어 퍼텐셜에너지의 제곱항에 대응하는 파인먼 다이어그램은 다음과 같다.

$t_i < t_k < t_l < t_f$

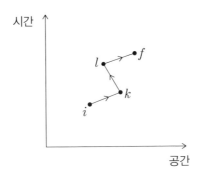

이렇게 τ에 대해 계속 전개하면 다음 식을 얻는다.

$$K(f|i) = K_0(f|i) + K_1(f|i) + K_2(f|i) + \cdots$$

이런 식으로 전자가 여러 번 힘을 받는 경우의 전파인자를 계산할 수 있고, 이것은 파인먼 다이어그램으로 간단하게 나타낼 수 있다.

네 번째 만남

양자전기역학

양자장론을 만든 물리학자들 _ 포크, 요르단, 위그너

정교수   양자전기역학을 이해하려면 양자장론을 먼저 알아야 해.

물리군   양자역학과는 다른 이론인가요?

정교수   양자역학은 위치와 운동량의 불확정성을 동시에 0으로 만들 수 없다는 이론이야. 그에 비해 양자장론은 입자의 생성과 소멸에 대한 이론이지. 양자장론을 수학으로 설명하는 것은 조금 뒤로 미루고 먼저 양자장론을 만든 과학자들의 이야기를 해볼게.

양자장론은 포크, 디랙, 요르단, 위그너, 페르미, 파울리와 같은 쟁쟁한 물리학자들에 의해 만들어졌어. 디랙과 페르미, 파울리에 대해서는 이 시리즈의 다른 책에서 이야기했으니까, 여기에서는 포크, 요르단, 위그너를 소개하려고 해. 첫 번째 물리학자는 포크야.

포크(Vladimir Aleksandrovich Fock, 1898~1974.
사진: 러시아 기념주화, 세 번째가 포크)

포크는 러시아 상트페테르부르크에서 태어났다. 그는 1922년에 페트로그라드 대학을 졸업한 후 대학원 과정을 밟고, 1932년에 그곳에서 교수가 되었다. 이후 바빌로프 주립 광학 연구소, 레닌그라드 물리 기술 연구소, 레베데프 물리 연구소에서 일했다.

포크의 주요 과학적 공헌은 양자물리학과 중력이론의 발전에 있으며, 역학, 이론 광학, 연속 매체 물리학 분야에도 크게 기여했다. 1926년 그는 포크 공간, 포크 표현, 포크 상태라는 아이디어를 냈고, 1930년에는 하트리-포크 근사법을 개발했다. 그는 또한 《암석 저항 연구 이론(1933)》이라는 책에서 지구물리학 탐사를 위한 전자기적 방법을 제시했다.

포크는 일반상대성이론과 다체문제에도 상당한 기여를 했다. 그는 과학적 근거를 바탕으로 아인슈타인의 일반상대성이론이 물리적 실체가 없다고 비판했다. 레닌그라드에서 포크는 이론물리학 과학 학교를 설립하고 그의 책을 통해 소련에서 물리학 교육을 발전시켰다.

이번에는 요르단에 대해 알아보자.

요르단(Ernst Pascual Jordan, 1902~1980)

요르단의 아버지 에른스트 요르단(Ernst Jordan)은 초상화와 풍경화로 유명한 화가로 하노버 공과대학의 미술 부교수였다. 성(姓)은 원래 요르다(Jorda)였으며 스페인 출신이었다.

요르단은 전통적인 종교 교육을 받으며 자랐다. 12세가 되었을 때 그는 성경의 문자적 해석과 다윈의 진화론을 조화시키려고 시도했다. 요르단의 종교 교사는 그에게 과학과 종교 사이에는 모순이 없다고 확신시켰고, 요르단은 일생에 걸쳐 과학과 종교의 관계에 대한 수많은 논문을 썼다.

1921년 요르단은 하노버 공과대학에 입학하여 동물학, 수학, 물리학을 공부했다. 당시 독일 대학생이 그랬듯이 그 또한 학위 취득 전에 다른 대학으로 학업을 옮겼다. 1923년 그는 괴팅겐 대학에서 수학자 힐베르트(David Hilbert)와 물리학자 조머펠트(Arnold Sommerfeld)에게 수학과 물리학을 배웠다.

괴팅겐에서 요르단은 한동안 수학자 쿠란트(Richard Courant)의 조수로 일했고, 그 후 막스 보른(Max Born) 밑에서 물리학을, 유전학자이자 인종 과학자인 알프레트 쿤(Alfred Kühn) 밑에서 유전을 공부하여 박사 학위를 받았다.

평생 언어장애로 어려움을 겪은 요르단은 준비되지 않은 연설을 할 때 종종 심하게 말을 더듬었다. 1926년 닐스 보어(Niels Bohr)는 치료비를 지불하겠다고 제안했고, 빌헬름 렌츠(Wilhelm Lenz)의 조언에 따라 요르단은 빈에 있는 알프레트 아들러(Alfred Adler)의 클리닉에서 치료를 받았다.

요르단은 막스 보른, 베르너 하이젠베르크(Werner Heisenberg)와 함께 양자역학에 관한 중요한 논문 시리즈의 공동 저자였다. 그는 제2차 세계대전 이전에 우주론으로 초점을 크게 전환하기 전에 초기 양자장론을 개척했다.

요르단은 양자역학 및 양자장론을 위한 관찰 가능한 대수를 만들기 위해 요르단 대수를 고안했다. 그 후 요르단 대수는 정수론, 복소해석, 최적화 및 기타 여러 순수 및 응용 수학 분야에 적용되었다.

독일의 제1차 세계대전 패배와 베르사유 조약은 요르단의 정치 신념에 지대한 영향을 미쳤다. 그는 민족주의적이고 우익적으로 변해 갔다. 1920년대 후반에는 공격적이고 호전적인 입장을 지지하는 수많은 기사를 썼다. 그는 반공주의자였으며 특히 러시아 혁명과 볼셰비키의 부상에 관심을 가졌다.

1933년 요르단은 필리프 레나르트, 요하네스 슈타르크와 같이 나치당에 가입했고, 나아가 SA 부대에도 들어갔다. 그는 나치의 민족주의와 반공주의를 지지했지만 동시에 '아인슈타인'과 다른 유대인 과학자들의 수호자로 남았다.

요르단은 자신이 새 정권에 영향력을 행사할 수 있기를 희망했다. 그의 프로젝트 중 하나는 아인슈타인으로 대표되는 현대 물리학과 코펜하겐의 새로운 양자 이론이 '볼셰비키의 유물론'에 대한 해독제가 될 수 있다고 나치를 설득하는 것이었다.

1939년 요르단은 독일 공군 부대인 루프트바페에 입대하여 한동안 페네뮌데 로켓 센터에서 기상 분석가로 일했다. 전쟁 중에는 나치당이 첨단 무기에 대한 다양한 계획을 세우는 데 일조했다.

요르단이 나치당에 가입하지 않았다면, 막스 보른과의 연구로 노벨 물리학상을 수상할 수 있었을 것이라고 많은 과학자들이 생각한다. 하지만 그는 나치당원을 고수했고 이로 인해 노벨 물리학상을 수상하지 못했다.

볼프강 파울리는 제2차 세계대전이 끝난 후 얼마 지나지 않아 요르단이 다시 학계에 취직할 수 있게 해주었다. 파울리의 도움으로 요르단은 1953년부터 함부르크 대학의 종신 교수로 임용되어 1971년에 명예교수가 될 때까지 재직했다.

이번에는 유진 위그너를 살펴보자.

위그너(Eugene Paul Wigner, 1902~1995, 1963년 노벨 물리학상 수상)

위그너는 1902년 오스트리아-헝가리 제국의 부다페스트에서 태어났다. 그의 부모는 중산층의 유대인이었다. 위그너는 9살이 될 때까지 전문 교사에게 홈스쿨링을 받았고, 초등학교 3학년 때부터 학교에 다니기 시작했다.

1915년부터 1919년까지 위그너는 아버지가 다녔던 학교인 부다페스트-파소리 에반젤리쿠스 김나지움에서 공부했다. 이곳에서 종교 교육은 의무였으며, 그는 랍비가 가르치는 유대교 수업에 참석했다. 위그너는 저명한 수학자 라츠(László Rátz)에게 수학을 배웠다. 이 학교 출신의 유명한 수학자로는 폰 노이만(John von Neumann)이 있다. 그는 위그너보다 한 살 아래였다.

파소리 김나지움(출처: Thaler Tamas/Wikimedia Commons)

1920년 김나지움을 졸업한 후, 위그너는 부다페스트 공과대학(Budapest University of Technical Sciences)에 입학했다. 1921년 그는 베를린 기술대학(Technische Hochschulein Berlin, 현재 Technische Universität Berlin)에 등록하여 화학공학을 공부했다. 그는 또한 독일 물리학 협회의 수요일 오후 콜로키아에 참석했다. 이

콜로키아에는 막스 플랑크, 막스 폰 라우에, 루돌프 라덴부르크, 베르너 하이젠베르크, 발터 네른스트, 볼프강 파울리, 알베르트 아인슈타인 등 주요 연구자들이 참여했다.

부다페스트 공과대학(출처: Varius/ Wikimedia Commons)

위그너는 카이저 빌헬름 물리화학 및 전기화학 연구소(현재 프리츠 하버 연구소)에서 일했다. 그곳에서 마이클 폴라니의 지도 아래 〈분자의 형성과 붕괴〉라는 제목의 박사 학위 논문을 완성했다.

부다페스트로 돌아온 위그너는 아버지의 무두질 공장에서 일하다가, 1926년 베를린의 카이저 빌헬름 연구소에서 카를 바이센베르크의 제안으로 X선 결정학 연구 조수가 되었다. 당시 위그너는 슈뢰딩거의 논문을 가지고 양자역학을 공부하기 시작했다.

그 후 위그너는 조머펠트의 부탁으로 괴팅겐 대학에서 힐베르트 교수의 조수로 일했다. 위그너는 양자역학에서 대칭 이론의 기초를 마련했으며 바일과 함께 군론을 양자역학에 처음 도입했다.

1930년 프린스턴 대학은 위그너를 1년 동안 강사로 채용했다. 그리고 그는 이 대학의 교수가 되었다. 1934년 프린스턴 대학에서 위그너는 여동생 마르기트(맨시)를 물리학자 폴 디랙에게 소개했고 두 사람은 결혼했다.

디랙과 그의 아내 마르기트
(출처: GFHund/Wikimedia Commons)

제2차 세계대전 동안 위그너는 맨해튼 프로젝트에서 우라늄을 무기급 플루토늄으로 변환할 수 있는 생산용 원자로를 설계하는 일을 맡았다. 1942년 7월, 그는 흑연 중성자 감속재와 수랭식을 갖춘 보수적인 100MW 원자로를 설계하는 데 성공했다.

생의 마지막에 이르러 위그너의 생각은 철학적으로 변했다. 1960년에 그는 〈자연과학에서 수학의 불합리한 효과〉라는 제목의 논문을 발표했다. 그는 생

1946년 미국 패터슨 장관과 위그너

네 번째 만남 _ 양자전기역학 165

물학과 인지가 우리 인간이 인식하는 물리적 개념의 기원이 될 수 있으며, 수학과 물리학이 그토록 잘 어울리는 행복한 우연의 일치는 '불합리'하고 설명하기 어려운 것처럼 보인다고 주장했다. 이 독창적인 논문은 다양한 학문 분야에서 큰 반향을 불러일으켰다.

양자장론 _ 입자의 생성과 소멸에 대한 이론

정교수 이제 양자장론에 대해 알아볼게. 양자장론은 1927년 디랙의 논문에서 처음 시작되었어. 영국의 디랙은 모든 입자가 숨어 있는 공간을 생각했지. 그는 이 숨어 있는 공간에 무한개의 입자가 있다고 보았네. 이 입자들이 우리가 관측할 수 있는 공간으로 튀어나오면 우리 눈에 보이게 돼. 이것을 입자의 생성이라고 불러. 그리고 우리가 관측할 수 있는 공간에서 숨어 있는 공간으로 입자가 들어가면 사라지는데 이것을 입자의 소멸이라고 한다네. 그런데 입자는 크게 두 종류로 나눌 수 있어.

물리군 어떻게요?

정교수 하나의 양자 상태에 무한개까지 채워지는 입자를 보존이라 하고, 하나의 양자 상태에 한 개[6]까지만 채워지는 입자를 페르미온이라고 부르지.

---

[6] 페르미온의 스핀을 고려하지 않는 경우를 생각한다. 만일 페르미온의 스핀을 고려하면 스핀이 서로 반대인 두 개의 입자가 하나의 양자 상태에 들어올 수 있다.

물리군   어떤 입자가 보존이고 어떤 입자가 페르미온이죠?

정교수   페르미온은 물질을 구성하는 입자라네. 전자, 양성자, 중성자와 같은 입자가 페르미온이지. 보존은 힘을 매개하는 캐치볼 입자야. 전자기힘을 매개하는 광자나 약력을 매개하는 W입자, Z입자와 같은 것들이 보존일세.

물리군   그렇군요.

정교수   그럼 대표적인 보존 입자인 광자를 생각해 볼까? 우리가 관측할 수 있는 공간에 광자가 한 개도 없는 경우를 상상해 보게. 디랙은 이러한 상태를 진공 상태라고 불렀어.

디랙은 우리가 관측할 수 있는 공간에 광자의 개수가 n개인 상태를 나타내는 켓벡터를

$$|n\rangle$$

이라고 쓰기로 했다. 그러니까 진공 상태를 나타내는 켓벡터는

$$|0\rangle$$

이라고 쓰면 될 것이다.

여기서 0은 광자가 0개라는 뜻이다. 이 상태는 정규화된 상태로 약속한다. 다시 말해 이 켓벡터의 크기는 1이다. 이것을 수식으로 쓰면

$$\langle 0|0\rangle = \||0\rangle\|^2 = 1$$

이다. 일반적으로 광자수가 n개인 상태도 정규화된 상태로 약속한다. 즉,

$$\langle n | n \rangle = \| | n \rangle \|^2 = 1 \qquad (4\text{-}2\text{-}1)$$

이 된다.

디랙은 우리가 관측할 수 있는 공간에서 광자수를 하나 감소시키는 연산자를 $\hat{a}$로 쓰고 소멸 연산자라고 불렀고, 광자수를 하나 증가시키는 연산자를 $\hat{a}^\dagger$로 쓰고 생성 연산자라고 불렀다. 여기서 $\hat{a}^\dagger$는 $\hat{a}$의 '수반'이다. 그러므로

$$(\hat{a}^\dagger)^\dagger = \hat{a}$$

가 된다.

물리군 생성 연산자는 숨어 있는 공간 속의 광자 한 개를 관찰 가능한 공간으로 보내는 역할을 하고, 소멸 연산자는 관찰 가능한 공간의 광자 한 개를 숨어 있는 공간으로 보내는 역할을 하는군요.

정교수 맞아. 디랙은 광자와 같은 보존은 다음 관계를 만족한다고 생각했지.

$$\hat{a}\hat{a}^\dagger - \hat{a}^\dagger\hat{a} = I \qquad (4\text{-}2\text{-}2)$$

여기서 I는 항등 연산자야. 따라서

$$I | n \rangle = | n \rangle \qquad (4\text{-}2\text{-}3)$$

이 되지.

그러니까 진공 상태에 $\hat{a}^\dagger$를 작용하면 광자 한 개가 생성되고, 광자가 한 개인 상태에 $\hat{a}^\dagger$를 작용하면 광자 두 개인 상태가 되고, 광자가 두 개인 상태에 $\hat{a}^\dagger$를 작용하면 광자 세 개인 상태가 된다. 그러므로

$$\hat{a}^\dagger |0\rangle \to |1\rangle$$

$$\hat{a}^\dagger |1\rangle \to |2\rangle$$

$$\hat{a}^\dagger |2\rangle \to |3\rangle$$

과 같다. 일반적으로 광자가 n개인 상태에 $\hat{a}^\dagger$를 작용하면 광자가 $(n+1)$개인 상태가 되므로

$$\hat{a}^\dagger |n\rangle \to |n+1\rangle$$

로 나타낼 수 있다. 이제 다음과 같이 비례상수를 도입하자.

$$\hat{a}^\dagger |n\rangle = c_{n+1} |n+1\rangle \qquad (4\text{-}2\text{-}4)$$

따라서

$$\hat{a}^\dagger |0\rangle = c_1 |1\rangle \qquad (4\text{-}2\text{-}5)$$

$$\hat{a}^\dagger |1\rangle = c_2 |2\rangle \qquad (4\text{-}2\text{-}6)$$

$$\hat{a}^\dagger |2\rangle = c_3 |3\rangle \qquad (4\text{-}2\text{-}7)$$

과 같다. 여기서 $c_1, c_2, c_3, \cdots$은 실수이다.

이번에는 광자수가 줄어드는 경우를 생각해 보자.

광자가 한 개인 상태에 $\hat{a}$를 작용하면 광자 0개인 상태(진공 상태)가 되고, 광자가 두 개인 상태에 $\hat{a}$를 작용하면 광자 한 개인 상태가 되고, 광자가 세 개인 상태에 $\hat{a}$를 작용하면 광자 두 개인 상태가 된다. 그러니까

$$\hat{a}|1\rangle \to |0\rangle$$

$$\hat{a}|2\rangle \to |1\rangle$$

$$\hat{a}|3\rangle \to |2\rangle$$

와 같다. 일반적으로 광자가 n개인 상태에 $\hat{a}$를 작용하면 광자가 $(n-1)$개인 상태가 되므로

$$\hat{a}|n\rangle \to |n-1\rangle$$

로 나타낼 수 있다. 이제 다음과 같이 비례상수를 도입하자.

$$\hat{a}|n\rangle = d_n |n-1\rangle \qquad (4\text{-}2\text{-}8)$$

따라서

$$\hat{a}\,|\,1\,\rangle = d_1\,|\,0\,\rangle \qquad (4\text{-}2\text{-}9)$$

$$\hat{a}\,|\,2\,\rangle = d_2\,|\,1\,\rangle \qquad (4\text{-}2\text{-}10)$$

$$\hat{a}\,|\,3\,\rangle = d_3\,|\,2\,\rangle \qquad (4\text{-}2\text{-}11)$$

와 같다. 여기서 $d_1, d_2, d_3, \cdots$은 실수이다.

물리군 $c_1, c_2, c_3, \cdots$과 $d_1, d_2, d_3, \cdots$은 어떻게 결정하죠?
정교수 식 (4-2-5)를 보게.

$$\hat{a}^\dagger\,|\,0\,\rangle = c_1\,|\,1\,\rangle$$

양변의 왼쪽에 $\langle 1|$을 작용하면

$$\langle 1\,|\,\hat{a}^\dagger\,|\,0\,\rangle = c_1$$

이 돼. 한편 식 (4-2-9)를 볼까?

$$\hat{a}\,|\,1\,\rangle = d_1\,|\,0\,\rangle$$

이 식 양변의 왼쪽에 $\langle 0|$을 작용하면

$$\langle 0\,|\,\hat{a}\,|\,1\,\rangle = d_1$$

이 돼. 그런데

$$d_1 = (\langle 0 | \cdot \hat{a} \cdot | 1 \rangle)^\dagger$$

$$= (|1\rangle)^\dagger \cdot (\hat{a})^\dagger \cdot (\langle 0|)^\dagger$$

$$= \langle 1 | \cdot (\hat{a})^\dagger \cdot | 0 \rangle$$

$$= \langle 1 | (\hat{a})^\dagger | 0 \rangle$$

$$= c_1$$

이야. 일반적으로

$$d_n = c_n \tag{4-2-12}$$

이 되지. 따라서 다음과 같이 쓸 수 있어.

$$\hat{a} | n \rangle = d_n | n-1 \rangle \tag{4-2-13}$$

$$\hat{a}^\dagger | n \rangle = d_{n+1} | n+1 \rangle \tag{4-2-14}$$

물리군 진공 상태에 $\hat{a}$를 작용하면 어떻게 되나요?
정교수 그런 상태는 존재하지 않아. 그러니까

$$\hat{a} | 0 \rangle = 0$$

이라고 두어야 해.

물리군 그렇다면 $d_0 = 0$이 되는군요.

정교수 맞아.

물리군 이제 d_n을 구하는 일만 남았네요.

정교수 그렇지. 식 (4-2-2)에 $|n\rangle$을 작용하면

$$\hat{a}\hat{a}^\dagger|n\rangle - \hat{a}^\dagger\hat{a}|n\rangle = I|n\rangle \qquad (4\text{-}2\text{-}15)$$

이고,

$$\hat{a}\hat{a}^\dagger|n\rangle = \hat{a}(\hat{a}^\dagger|n\rangle)$$

$$= \hat{a}(d_{n+1}|n+1\rangle)$$

$$= d_{n+1}(\hat{a}|n+1\rangle)$$

$$= d_{n+1}(d_{n+1}|n\rangle)$$

$$= d_{n+1}^2|n\rangle$$

이 돼. 마찬가지로

$$\hat{a}^\dagger\hat{a}|n\rangle = d_n^2|n\rangle$$

이지. 그러니까 식 (4-2-15)는

$$d_{n+1}^2 - d_n^2 = 1$$

이야. 이 식은

$$d_n^2 = n$$

이면 만족해. 즉,

$$d_n = \sqrt{n}$$

이 돼. 따라서 식 (4-2-13)과 (4-2-14)는 다음과 같이 쓸 수 있어.

$$\hat{a}\,|\,n\,\rangle = \sqrt{n}\,|\,n-1\,\rangle \qquad (4\text{-}2\text{-}16)$$

$$\hat{a}^\dagger\,|\,n\,\rangle = \sqrt{n+1}\,|\,n+1\,\rangle \qquad (4\text{-}2\text{-}17)$$

여기서 $\hat{N} = \hat{a}^\dagger \hat{a}$ 라고 하면

$$\hat{N}\,|\,n\,\rangle = n\,|\,n\,\rangle \qquad (4\text{-}2\text{-}18)$$

이야. 그러니까 광자가 n개인 상태를 나타내는 켓벡터에 $\hat{N}$을 작용하면 광자수와 광자가 n개인 상태를 나타내는 켓벡터의 곱이 되지. 그래서 $\hat{N}$을 수 연산자라고 부른다네. 우리가 광자를 다루고 있으니까 이 수 연산자는 광자수 연산자이지. 즉, 수 연산자의 고윳값이 광자수일세.

물리군 광자는 보존이니까 하나의 상태에 무한히 많은 광자가 있을 수 있잖아요? 그것도 증명할 수 있나요?

정교수 물론이야. 식 (4-2-2)에서 왼쪽에 $\langle n|$을, 오른쪽에 $|n\rangle$을 작용하면

$$\langle n | \hat{a}\hat{a}^\dagger | n \rangle - \langle n | \hat{a}^\dagger \hat{a} | n \rangle = \langle n | I | n \rangle \qquad (4\text{-}2\text{-}19)$$

이야. 즉,

$$\langle n | \hat{a}\hat{a}^\dagger | n \rangle = n + 1$$

이 되지. 그런데

$$(\hat{a}^\dagger | n \rangle)^\dagger = \langle n | a$$

이니까

$$\langle n | \hat{a}\hat{a}^\dagger | n \rangle = (\hat{a}^\dagger | n \rangle)^\dagger (\hat{a}^\dagger | n \rangle) = \| \hat{a}^\dagger | n \rangle \|^2 \qquad (4\text{-}2\text{-}20)$$

이 돼. 벡터의 크기의 제곱은 음수가 아니므로

$$\| \hat{a}^\dagger | n \rangle \|^2 \geq 0$$

이야. 그러니까

$$n + 1 \geq 0$$

이 돼. 이 식을 만족하는 n의 값은 다음과 같아.

$$n = 0, 1, 2, 3, \cdots$$

따라서 하나의 상태에 들어갈 수 있는 광자수는 무한개까지 가능하지. 바로 보존 입자의 특징이야.

물리군 그렇군요. 그럼 스핀을 고려하지 않을 때 페르미온 입자는 왜 하나의 상태에 두 개 이상의 입자가 못 들어가나요?

정교수 페르미온 입자의 생성 연산자 $\hat{b}^\dagger$와 소멸 연산자 $\hat{b}$는 보존 입자와는 다른 관계를 만족해. 이들 사이의 관계는 다음과 같아.

$$\hat{b}\hat{b}^\dagger + \hat{b}^\dagger\hat{b} = I \qquad (4\text{-}2\text{-}21)$$

물리군 뺄셈이 덧셈으로 바뀌었군요.

정교수 맞아. 그러니까 페르미온 입자의 수 연산자는

$$\hat{N} = \hat{b}^\dagger\hat{b} \qquad (4\text{-}2\text{-}22)$$

이고, 페르미온 입자가 n개인 상태를 나타내는 켓벡터 $|n\rangle$은

$$\hat{N}|n\rangle = n|n\rangle$$

이 되고,

$$\hat{b}^\dagger|n\rangle = \sqrt{n+1}\,|n+1\rangle \qquad (4\text{-}2\text{-}23)$$

$$\hat{b}|n\rangle = \sqrt{n}\,|n\rangle \qquad (4\text{-}2\text{-}24)$$

이 되지. 식 (4-2-21)에서 왼쪽에 $\langle n|$을, 오른쪽에 $|n\rangle$을 작용하면

$$\langle n|\hat{b}\hat{b}^\dagger|n\rangle + \langle n|\hat{b}^\dagger\hat{b}|n\rangle = \langle n|I|n\rangle \qquad (4\text{-}2\text{-}25)$$

이야. 즉,

$$\langle n | \hat{b}\hat{b}^\dagger | n \rangle = 1 - n$$

이 되지. 그런데

$$\langle n | \hat{b}\hat{b}^\dagger | n \rangle = \| \hat{b}^\dagger | n \rangle \|^2 \geq 0$$

이므로

$$1 - n \geq 0$$

이지. 이 식을 만족하는 n의 값은 다음과 같아.

$$n = 0, 1$$

그러니까 하나의 상태에 들어갈 수 있는 페르미온 입자의 수는 한 개까지 가능하지. 바로 페르미온 입자의 특징이야.

양자전기역학의 창시자 _ 슈윙거와 도모나가 신이치로

정교수 양자전기역학은 슈윙거, 파인먼, 도모나가 신이치로에 의해 독립적으로 연구된 이론이야. 이 세 사람은 노벨 물리학상을 공동 수상하지. 파인먼에 대해서는 앞에서 얘기했으니까 나머지 두 사람을 소개할게. 먼저 슈윙거를 알아보세.

슈윙거(Julian Seymour Schwinger, 1918~1994, 1965년 노벨 물리학상 수상)

슈윙거는 미국 뉴욕시에서 태어났다. 그의 아버지와 어머니는 유대인으로 모두 잘나가는 의류 제조업자였지만, 1929년 월 스트리트 붕괴 이후 사업이 쇠퇴했다. 가족은 정통 유대인의 전통을 따랐다.

1932년부터 1934년까지 슈윙거는 당시 영재 학생들을 위한 학교로 높은 평가를 받던 타운센드 해리스 고등학교에 다녔다. 고등학생 시절 그는 뉴욕 시립 대학(CCNY) 도서관에서 디랙의 논문을 읽을 정도로 수학과 물리학에 특출난 재능을 보였다.

타운센드 해리스 고등학교 (출처: Jellybean100/ Wikimedia Commons)

1934년 가을, 슈윙거는 뉴욕 시립 대학에 입학했다. 그는 다른 학생들보다 수학과 물리학을 잘한 덕에 수업에 제대로 들어가지 않았음에도 수학과 물리학에서 매우 좋은 성적을 거두었다. 하지만 영어 등 다른 과목은 그다지 성적이 좋지 않았다.

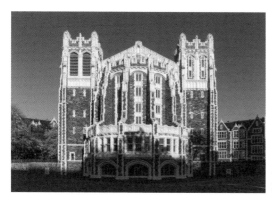

뉴욕 시립 대학(출처: King of Hearts/Wikimedia Commons)

뉴욕 시립 대학을 졸업한 슈윙거는 컬럼비아 대학에서 대학원 과정을 밟았다. 물리학과 수학을 제외한 다른 과목 성적이 좋지 않았던 그는 컬럼비아 대학에서 장학금을 받을 수 없는 상황이었다. 컬럼비아 대학 물리학과 강사인 로이드 모츠(Lloyd Motz)는 슈윙거의 재능을 알아보고 라비(Isidor Isaac Rabi, 1944년 노벨 물리학상 수상) 교수에게 장학금을 요청했다. 라비 교수는 한스 베테 교수에게 슈윙거가 쓴 양자전기역학에 관한 미발표 논문을 보여주었고, 이를 본 베테 교수가 슈윙거에게 장학금을 주기로 결정했다.

슈윙거는 낮에는 주로 잠을 자고 밤을 새워 연구했다. 그는 1939년 21세의 나이로 라비 교수의 지도하에 박사 학위를 취득했다.

1939년 가을, 슈윙거는 캘리포니아 대학교 버클리의 로버트 오펜하이머 밑에서 연구를 했고, 그곳에서 연구원으로 2년 동안 머물렀다.

   1941년 슈윙거는 퍼듀 대학 교수가 되었다. 제2차 세계대전 당시 그는 MIT 방사선 연구소에서 레이더 개발에 대한 이론적 지원을 제공했다. 전쟁이 끝난 후에는 퍼듀를 떠나 하버드 대학으로 가서 1945년부터 1974년까지 교수 생활을 했다.

   슈윙거는 그린 함수를 이용해 양자장론을 만들었다. 이를 통해 그는 양자전기역학에서 전자의 자기모멘트에 대한 첫 번째 수정 사항을 명확하게 계산할 수 있었다. 그리고 전기장 속에서 양자 터널 효과에 의해 전자─양전자 쌍이 생성되는 속도를 계산했다.

   또한 슈윙거는 뉴트리노가 전자 뉴트리노와 뮤온 뉴트리노 등 여러 종류가 있을 수 있다는 것을 알아냈다. 1960년대에 그는 현재 슈윙거 모델로 알려진 시공간에서의 양자전기역학을 공식화했다. 슈윙거는 양자전기역학(QED) 연구로 1965년 리처드 파인먼, 도모나가 신이치로와 함께 노벨 물리학상을 공동 수상했다.

   슈윙거는 73명의 박사 학위 논문을 지도했는데, 그의 학생 중 4명이 노벨상을 수상했다. 노벨상 수상자는 글라우버(Roy Glauber, 물리), 모텔손(Ben Roy Mottelson, 물리), 글래쇼(Sheldon Glashow, 물리) 및 월터 콘(Walter Kohn, 화학)이다.

   유명한 물리학자로서 슈윙거는 종종 동시대의 또 다른 전설적인 물리학자인 리처드 파인먼과 비교되었다. 슈윙거는 파인먼 다이어그램을 이용하면 학생들이 입자 물리학을 제대로 이해할 수 없다고 여겨 수업에서 파인먼 다이어그램 사용을 금지했다.

이번에는 일본의 도모나가 신이치로에 대해 살펴보자.

도모나가 신이치로(朝永振一郎, 1906~1979, 1965년 노벨 물리학상 수상)

도모나가 신이치로는 1906년 일본 도쿄에서 태어났다. 그는 일본 철학자 도모나가 산주로(朝永三十郎)의 둘째이자 장남이었다. 도모나가 신이치로는 1926년 교토 대학에 입학했다. 노벨상 수상자이기도 한 유카와 히데키는 학부 시절 그의 동급생 중 한 명이었다.

교토 대학 대학원 재학 중 도모나가 신이치로는 3년 동안 대학에서 조교로 일했다. 1931년 대학원 졸업 후 이화학연구소의 니시나 그룹에 합류했다. 1937년에는 라이프치히 대학에서 근무하면서 하이젠베르크의 연구 그룹과 협력했다. 2년 후 제2차 세계대전이 발발하여 일본으로 돌아갔으나, 라이프치히에서 수행한 연구 논문으로 핵물질 연구에 관한 박사 학위를 도쿄 대학에서 받았다.

도모나가 신이치로는 박사 학위 취득 후 도쿄 교육대학(현재 츠쿠바 대학) 교수로 임명되었다. 1948년에 그는 양자전기역학에서 발생하는 무한한 양이 서로 상쇄됨을 보이려고 시도했다가 실패한 시드니 댄코프(Sidney Dancoff)의 1939년 논문을 재검토했다. 도모나가 신이치로는 자신의 초다시간 이론을 이용해 댄코프의 논문에서 틀린 점을 발견했고, 양자전기역학에서 발생하는 무한한 양이 서로 상쇄될 수 있음을 보였다.

1949년에 도모나가 신이치로는 오펜하이머의 초청을 받아 프린스턴 고등연구소에서 일했다. 그는 양자역학계의 집단 진동에 관한 다체문제를 연구했다. 1955년에는 도쿄 대학 핵 연구소 설립을 주도했다. 그 후 1965년에 양자전기역학 연구로 슈윙거, 파인먼과 함께 노벨 물리학상을 수상했다.

양자전기역학 _ 파인먼 다이어그램을 이용하여

정교수 이제 양자전기역학 내용과 그것으로 어떤 성공을 거두었는지를 알아볼 차례야. 수식으로는 너무 복잡하니까 파인먼 다이어그램을 통해 설명할게.[7]

물리군 그림으로 배우는 게 이해하기 쉽죠.

---

7) 양자전기역학은 물리학과 대학원 소립자물리학 전공자가 석사 과정에서 배우는 과목이다.

정교수 고전전기역학에서는 빛을 전자기파라는 파동으로 다뤄. 반면에 양자전기역학에서는 빛을 광자라는 이름의 양자로 다룬다네. 즉, 양자의 세계에서 모든 입자는 양자가 돼. 전자도 양자이지. 양자전기역학은 두 종류의 양자―광자와 전자―의 상호작용을 다루는 물리학이야.

파인먼 다이어그램은 가로축을 공간축으로, 세로축을 시간축으로 택한다. 시간축에서는 아래가 과거이고 위가 미래이다. 여기서 한 점은 시간과 공간을 나타내므로 시공간의 한 점이 된다.

전자가 시공간상의 한 점 1에서 다른 점 2로 가는 경우 파인먼 다이어그램은 다음 그림과 같다.

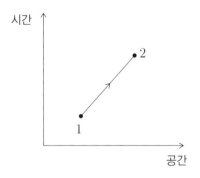

이 그림에서는 1에서 2로 가는 직선으로 표현했지만 사실 1에서 2로 가는 많은 경로를 함께 고려해야 한다. 예를 들어 다음과 같은 경로가 있을 수 있다.

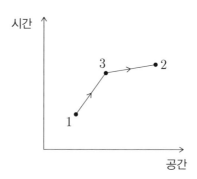

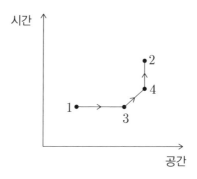

이러한 모든 경로를 고려한 전파인자를

$K(2\,|\,1)$

로 나타낸다.

한편 광자가 시공간상의 한 점 1에서 다른 점 2로 가는 경우의 파인먼 다이어그램은 다음 그림과 같이 물결선으로 나타낸다.

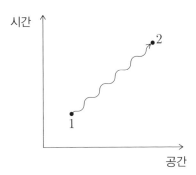

이제 전자가 1에서 2로 가는데 중간에 3에서 광자를 방출하는 경우를 파인먼 다이어그램으로 그려보자.

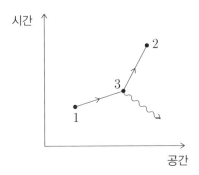

마찬가지로 전자가 1에서 2로 가는데 중간에 3에서 광자를 흡수하는 경우를 파인먼 다이어그램으로 그리면 다음과 같다.

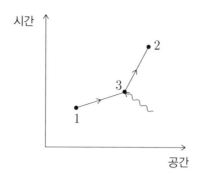

물리군 그림으로 보니까 이해가 잘되네요.

정교수 이번에는 1, 2에 있는 전자가 3, 4로 이동하는 경우를 볼까? 이때 가능한 경우는 1에 있는 전자가 3으로 가고 2에 있는 전자가 4로 가거나, 또는 1에 있는 전자가 4로 가고 2에 있는 전자가 3으로 가는 것이지. 첫 번째 경우를 파인먼 다이어그램으로 그리면 다음과 같아.

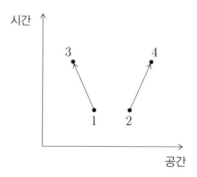

두 번째 경우를 파인먼 다이어그램으로 그리면 다음과 같지.

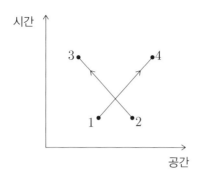

물리군 또 다른 경우도 있나요?

정교수 물론이야. 1에서 3으로 전자가 곧장 가지 않고 5에서 광자를 방출하는 걸세. 2에서 4로 가는 전자가 6에서 이 광자를 받는 거지. 이것을 파인먼 다이어그램으로 그리면 다음과 같아.

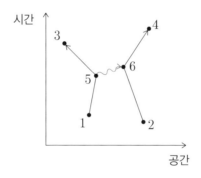

또 다른 경우로 1에서 4로 전자가 곧장 가지 않고 6에서 광자를 방출하는 것도 생각할 수 있어. 이때 2에서 3으로 가는 전자가 5에서 이 광

자를 받는 거지. 이것을 파인먼 다이어그램으로 그리면 다음과 같아.

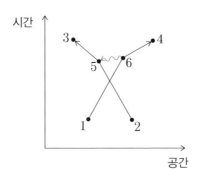

물리군 광자 하나를 주고받는군요.

정교수 맞아. 다음 그림과 같이 두 개의 광자를 주고받는 경우도 있어.

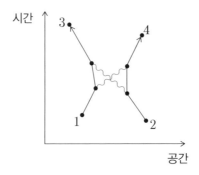

이런 식으로 세 개, 네 개, 다섯 개 등의 광자를 주고받는 경우를 모두 고려해야 해. 그러니까 모든 서로 다른 파인먼 다이어그램에서 계산되는 확률진폭을 더해야 전체 확률진폭을 결정할 수 있어.

(전체 확률진폭)

=(모든 서로 다른 파인먼 다이어그램에서의 확률진폭의 합)

파인먼 다이어그램은 전자와 양전자가 만나 빛(광자)으로 소멸했다가 이 빛이 다시 전자와 양전자 쌍을 만드는 과정도 나타낼 수 있다네. 전자를 e^-, 양전자를 e^+라고 하면 다음 그림과 같아.

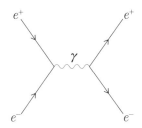

위 그림에서 γ는 감마선을 이루는 광자를 나타내지.

물리군 양자전기역학의 실험적인 증거가 있나요?

정교수 물론이야. 바로 전자의 자기모멘트 계산이라네.

물리군 전자의 자기모멘트가 뭐죠?

정교수 외부에서 전자에 일정한 자기장 B를 걸어주면 전자가 받는 자기력에 대한 퍼텐셜에너지의 크기는 자기장의 세기에 비례해. 그 비례상수를 μ로 쓰고 이것을 자기모멘트라고 불러.

물리군 그러니까 원자 속의 전자에 자석을 가져다 대면 되겠네요.

정교수 그렇지. 자석에서는 일정한 자기장이 나오니까 말일세. 이때

전자의 자기모멘트는 양자역학에 의해

$$\mu = g\left(\frac{\hbar e}{4m}\right)$$

로 알려졌어. 이때 g를 g-인자라고 부르는데 슈뢰딩거 방정식을 풀어서 구하면

$g = 2$　(양자역학 이론)

가 돼. 그런데 실험 결과는 2보다 조금 큰 값이 나왔어.

$g = 2.00231930482$　(실험)

물리군　거의 비슷하네요.

정교수　물론이지. 이것은 다음 그림과 같은 파인먼 다이어그램만 고려한 결과야.

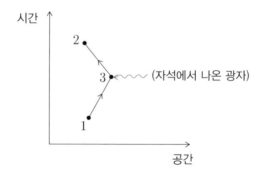

그러니까 모든 파인먼 다이어그램을 고려해서 계산하면 실험값과 좀 더 일치하는 결과를 찾을 수 있지.

물리군 어떤 다이어그램을 추가할 수 있나요?

정교수 다음 그림을 보게.

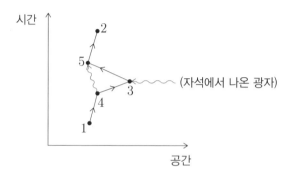

위 그림은 전자가 1에서 출발해 4에서 광자를 5로 방출하고, 3에서 자석에 의한 광자를 흡수해 5로 가서 4에서 방출된 광자를 흡수한 후 2로 가는 파인먼 다이어그램이야. 이런 식으로 모든 가능한 파인먼 다이어그램을 고려하여 계산한 g-인자는

$g = 2.00231930076$ (양자전기역학)

이지.

물리군 와우! 실험값과 거의 일치하네요.

정교수 이것이 바로 양자전기역학의 최대 성공 업적이야. 그래서 이 이론을 만든 세 사람에게 노벨 물리학상을 수여한 거라네.

만남에 덧붙여

A NEW NOTATION FOR QUANTUM MECHANICS

By P. A. M. DIRAC

Received 29 April 1939

In mathematical theories the question of notation, while not of primary importance, is yet worthy of careful consideration, since a good notation can be of great value in helping the development of a theory, by making it easy to write down those quantities or combinations of quantities that are important, and difficult or impossible to write down those that are unimportant. The summation convention in tensor analysis is an example, illustrating how specially appropriate a notation can be.

The notation in current use in quantum mechanics is fairly well suited to its purposes, but has some drawbacks. One has to deal with vectors in Hilbert space, representing the states of a dynamical system, and with linear operators, representing dynamical variables, and one sometimes makes calculations using the vectors and linear operators directly, treating them as abstract quantities which can be combined together algebraically according to certain rules, while at other times one works with coordinates (or *representatives*, as they are called) of these quantities. For the two styles of calculation two distinct notations are used, which do not fit together very naturally and which give rise to an awkward jump in the flow of one's thoughts when one changes from one to the other. In the present note a new notation is set up, which provides a neat and concise way of writing, in a single scheme, both the abstract quantities themselves and their coordinates, and thus leads to a unification of ideas.

A Hilbert-space vector, which was denoted in the old notation by the letter ψ, will now be denoted by a special new symbol $\rangle$. If we are concerned with a particular vector, specified by a label, a say, which would be used as a suffix to the ψ in the old notation, we write it $|a\rangle$. It may be that the label is very complicated, consisting of many letters, but we can always write down the vector conveniently in the new notation, simply by enclosing the label between $|$ on the left and $\rangle$ on the right. We have also to deal with another kind of Hilbert-space vector, the conjugate imaginary of the first kind. This was denoted in the old notation by ϕ or $\bar{\psi}$, and will now be denoted by $\langle$. If one of them is specified by a label a, we write it $\langle a|$.

A pair of vectors, one of each kind, have a symbolic product, which is a number. In the old notation such a product was denoted by $\phi\psi$, $\phi\psi_a$, $\phi_a\psi$, or $\phi_a\psi_b$, according as the vectors have labels or not. In the new notation these products will be denoted by $\langle\ \rangle$, $\langle|a\rangle$, $\langle a|\rangle$ and $\langle a|b\rangle$ respectively. Note that in the last of these it is not necessary to have the | occurring twice, as a simple juxtaposition of $\langle a|$ and $|b\rangle$ would give.

Let us now introduce a representation, say the one in which each of a certain complete set of commuting observables q is diagonal. This gives us a set of basic vectors in the Hilbert space, one for each set of eigenvalues q' for the q's. These basic vectors, which were denoted in the old notation by $\psi(q')$, will now be denoted by $|q'\rangle$, the q' being treated as an ordinary label. Similarly, the conjugate imaginary basic vectors, which were denoted in the old notation by $\phi(q')$, will now be denoted by $\langle q'|$. To get the representative of any vector $|a\rangle$, we must form its product with the basic vector $\langle q'|$, which gives us $\langle q'|a\rangle$. This is similar to the bracket expression $(q'|a)$ for the representative of ψ_a in the old notation. *But while the bracket expression $(q'|a)$ of the old notation has to be introduced as an extraneous symbol, independent of what has gone before, the $\langle q'|a\rangle$ of the new notation appears naturally as a symbolic product.*

This illustrates the main feature of the new notation. With the old notation there are often two quite different ways of writing a quantity, one as a product of abstract symbols and the other by means of the bracket notation—for example, the above bracket expression $(q'|a)$ could also be written as the product $\phi(q')\,\psi_a$—but in the new notation there is always only one, and complete continuity is preserved when one passes from expressions involving abstract symbols to representatives.

This feature is illustrated also by transformation functions. If we take a second representation in which, say, each of the complete set of commuting observables ξ is diagonal, the transformation function would be written in the old notation either as the bracket expression $(q'|\xi')$ or as the symbolic product of two basic vectors $\phi(q')\,\psi(\xi')$. In the new notation these two ways of writing it coalesce into $\langle q'|\xi'\rangle$.

The development of the new notation to include linear operators and observables can be effected without difficulty. Below is a list of the various types of quantity involving a linear operator α, written on the left in the old notation and on the right in the new.

$$\alpha\psi \qquad \alpha\rangle$$
$$\alpha\psi_a \qquad \alpha|a\rangle$$
$$\phi\alpha \qquad \langle\alpha$$
$$\phi_a\alpha \qquad \langle a|\alpha$$
$$\phi_a\alpha\psi \qquad \langle a|\alpha\rangle \text{ or } \langle a|\alpha|\rangle$$
$$\phi\alpha\psi_a \qquad \langle\alpha|a\rangle \text{ or } \langle|\alpha|a\rangle$$
$$\phi_a\alpha\psi_b \qquad \langle a|\alpha|b\rangle$$
$$\phi(q')\alpha\psi(q'') \text{ or } (q'|\alpha|q'') \qquad \langle q'|\alpha|q''\rangle.$$

The last of these is the representative of a linear operator and provides a further example of a quantity that can be written in two quite different ways with the old notation, but only in one with the new. Where two forms are given for writing an expression in the new notation, the former may be used for brevity when there is no danger of α being mistaken for the label of a vector in Hilbert space, otherwise the latter must be used.

Two general rules in connexion with the new notation may be noted, namely, *any quantity in brackets ⟨ ⟩ is a number*, and *any expression containing an unclosed bracket symbol ⟨ or ⟩ is a vector in Hilbert space, of the nature of a φ or ψ respectively*. As names for the new symbols ⟨ and ⟩ to be used in speech, I suggest the words *bra* and *ket* respectively.

St John's College
 Cambridge

논문 웹페이지

REVIEWS OF MODERN PHYSICS

Space-Time Approach to Non-Relativistic Quantum Mechanics

R. P. FEYNMAN

Cornell University, Ithaca, New York

Non-relativistic quantum mechanics is formulated here in a different way. It is, however, mathematically equivalent to the familiar formulation. In quantum mechanics the probability of an event which can happen in several different ways is the absolute square of a sum of complex contributions, one from each alternative way. The probability that a particle will be found to have a path $x(t)$ lying somewhere within a region of space time is the square of a sum of contributions, one from each path in the region. The contribution from a single path is postulated to be an exponential whose (imaginary) phase is the classical action (in units of $\hbar$) for the path in question. The total contribution from all paths reaching x, t from the past is the wave function $\psi(x, t)$. This is shown to satisfy Schroedinger's equation. The relation to matrix and operator algebra is discussed. Applications are indicated, in particular to eliminate the coordinates of the field oscillators from the equations of quantum electrodynamics.

1. INTRODUCTION

IT is a curious historical fact that modern quantum mechanics began with two quite different mathematical formulations: the differential equation of Schroedinger, and the matrix algebra of Heisenberg. The two, apparently dissimilar approaches, were proved to be mathematically equivalent. These two points of view were destined to complement one another and to be ultimately synthesized in Dirac's transformation theory.

This paper will describe what is essentially a third formulation of non-relativistic quantum theory. This formulation was suggested by some of Dirac's[1,2] remarks concerning the relation of classical action[3] to quantum mechanics. A probability amplitude is associated with an entire motion of a particle as a function of time, rather than simply with a position of the particle at a particular time.

The formulation is mathematically equivalent to the more usual formulations. There are, therefore, no fundamentally new results. However, there is a pleasure in recognizing old things from a new point of view. Also, there are problems for which the new point of view offers a distinct advantage. For example, if two systems A and B interact, the coordinates of one of the systems, say B, may be eliminated from the equations describing the motion of A. The inter-

---

[1] P. A. M. Dirac, *The Principles of Quantum Mechanics* (The Clarendon Press, Oxford, 1935), second edition, Section 33; also, Physik. Zeits. Sowjetunion **3**, 64 (1933).
[2] P. A. M. Dirac, Rev. Mod. Phys. **17**, 195 (1945).

[3] Throughout this paper the term "action" will be used for the time integral of the Lagrangian along a path. When this path is the one actually taken by a particle, moving classically, the integral should more properly be called Hamilton's first principle function.

action with B is represented by a change in the formula for the probability amplitude associated with a motion of A. It is analogous to the classical situation in which the effect of B can be represented by a change in the equations of motion of A (by the introduction of terms representing forces acting on A). In this way the coordinates of the transverse, as well as of the longitudinal field oscillators, may be eliminated from the equations of quantum electrodynamics.

In addition, there is always the hope that the new point of view will inspire an idea for the modification of present theories, a modification necessary to encompass present experiments.

We first discuss the general concept of the superposition of probability amplitudes in quantum mechanics. We then show how this concept can be directly extended to define a probability amplitude for any motion or path (position $vs.$ time) in space-time. The ordinary quantum mechanics is shown to result from the postulate that this probability amplitude has a phase proportional to the action, computed classically, for this path. This is true when the action is the time integral of a quadratic function of velocity. The relation to matrix and operator algebra is discussed in a way that stays as close to the language of the new formulation as possible. There is no practical advantage to this, but the formulae are very suggestive if a generalization to a wider class of action functionals is contemplated. Finally, we discuss applications of the formulation. As a particular illustration, we show how the coordinates of a harmonic oscillator may be eliminated from the equations of motion of a system with which it interacts. This can be extended directly for application to quantum electrodynamics. A formal extension which includes the effects of spin and relativity is described.

2. THE SUPERPOSITION OF PROBABILITY AMPLITUDES

The formulation to be presented contains as its essential idea the concept of a probability amplitude associated with a completely specified motion as a function of time. It is, therefore, worthwhile to review in detail the quantum-mechanical concept of the superposition of probability amplitudes. We shall examine the essential changes in physical outlook required by the transition from classical to quantum physics.

For this purpose, consider an imaginary experiment in which we can make three measurements successive in time: first of a quantity A, then of B, and then of C. There is really no need for these to be of different quantities, and it will do just as well if the example of three successive position measurements is kept in mind. Suppose that a is one of a number of possible results which could come from measurement A, b is a result that could arise from B, and c is a result possible from the third measurement C.[4] We shall assume that the measurements A, B, and C are the type of measurements that completely specify a state in the quantum-mechanical case. That is, for example, the state for which B has the value b is not degenerate.

It is well known that quantum mechanics deals with probabilities, but naturally this is not the whole picture. In order to exhibit, even more clearly, the relationship between classical and quantum theory, we could suppose that classically we are also dealing with probabilities but that all probabilities either are zero or one. A better alternative is to imagine in the classical case that the probabilities are in the sense of classical statistical mechanics (where, possibly, internal coordinates are not completely specified).

We define P_{ab} as the probability that if measurement A gave the result a, then measurement B will give the result b. Similarly, P_{bc} is the probability that if measurement B gives the result b, then measurement C gives c. Further, let P_{ac} be the chance that if A gives a, then C gives c. Finally, denote by P_{abc} the probability of all three, i.e., if A gives a, then B gives b, and C gives c. If the events between a and b are independent of those between b and c, then

$$P_{abc} = P_{ab}P_{bc}. \qquad (1)$$

This is true according to quantum mechanics when the statement that B is b is a complete specification of the state.

---

[4] For our discussion it is not important that certain values of a, b, or c might be excluded by quantum mechanics but not by classical mechanics. For simplicity, assume the values are the same for both but that the probability of certain values may be zero.

In any event, we expect the relation

$$P_{ac} = \sum_b P_{abc}. \quad (2)$$

This is because, if initially measurement A gives a and the system is later found to give the result c to measurement C, the quantity B must have had some value at the time intermediate to A and C. The probability that it was b is P_{abc}. We sum, or integrate, over all the mutually exclusive alternatives for b (symbolized by $\sum_b$).

Now, the essential difference between classical and quantum physics lies in Eq. (2). In classical mechanics it is always true. In quantum mechanics it is often false. We shall denote the quantum-mechanical probability that a measurement of C results in c when it follows a measurement of A giving a by $P_{ac}{}^q$. Equation (2) is replaced in quantum mechanics by this remarkable law:[5] There exist complex numbers φ_{ab}, φ_{bc}, φ_{ac} such that

$$P_{ab} = |\varphi_{ab}|^2, \quad P_{bc} = |\varphi_{bc}|^2, \text{ and } P_{ac}{}^q = |\varphi_{ac}|^2. \quad (3)$$

The classical law, obtained by combining (1) and (2),

$$P_{ac} = \sum_b P_{ab} P_{bc} \quad (4)$$

is replaced by

$$\varphi_{ac} = \sum_b \varphi_{ab} \varphi_{bc}. \quad (5)$$

If (5) is correct, ordinarily (4) is incorrect. The logical error made in deducing (4) consisted, of course, in assuming that to get from a to c the system had to go through a condition such that B had to have some definite value, b.

If an attempt is made to verify this, i.e., if B is measured between the experiments A and C, then formula (4) is, in fact, correct. More precisely, if the apparatus to measure B is set up and used, but no attempt is made to utilize the results of the B measurement in the sense that only the A to C correlation is recorded and studied, then (4) is correct. This is because the B measuring machine has done its job; if we wish, we could read the meters at any time without

disturbing the situation any further. The experiments which gave a and c can, therefore, be separated into groups depending on the value of b.

Looking at probability from a frequency point of view (4) simply results from the statement that in each experiment giving a and c, B had some value. The only way (4) could be wrong is the statement, "B had some value," must sometimes be meaningless. Noting that (5) replaces (4) only under the circumstance that we make no attempt to measure B, we are led to say that the statement, "B had some value," may be meaningless whenever we make no attempt to measure B.[6]

Hence, we have different results for the correlation of a and c, namely, Eq. (4) or Eq. (5), depending upon whether we do or do not attempt to measure B. No matter how subtly one tries, the attempt to measure B must disturb the system, at least enough to change the results from those given by (5) to those of (4).[7] That measurements do, in fact, cause the necessary disturbances, and that, essentially, (4) could be false was first clearly enunciated by Heisenberg in his uncertainty principle. The law (5) is a result of the work of Schroedinger, the statistical interpretation of Born and Jordan, and the transformation theory of Dirac.[8]

Equation (5) is a typical representation of the wave nature of matter. Here, the chance of finding a particle going from a to c through several different routes (values of b) may, if no attempt is made to determine the route, be represented as the square of a sum of several complex quantities—one for each available route.

---

[5] We have assumed b is a non-degenerate state, and that therefore (1) is true. Presumably, if in some generalization of quantum mechanics (1) were not true, even for pure states b, (2) could be expected to be replaced by: There are complex numbers φ_{abc} such that $P_{abc} = |\varphi_{abc}|^2$. The analog of (5) is then $\varphi_{ac} = \sum_b \varphi_{abc}$.

[6] It does not help to point out that we *could* have measured B had we wished. The fact is that we did not.
[7] How (4) actually results from (5) when measurements disturb the system has been studied particularly by J. von Neumann (*Mathematische Grundlagen der Quantenmechanik* (Dover Publications, New York, 1943)). The effect of perturbation of the measuring equipment is effectively to change the phase of the interfering components, by θ_b, say, so that (5) becomes $\varphi_{ac} = \sum_b e^{i\theta_b} \varphi_{ab} \varphi_{bc}$. However, as von Neumann shows, the phase shifts must remain unknown if B is measured so that the resulting probability P_{ac} is the square of φ_{ac} averaged over all phases, θ_b. This results in (4).
[8] If **A** and **B** are the operators corresponding to measurements A and B, and if ψ_a and ψ_b are solutions of $\mathbf{A}\psi_a = a\psi_a$ and $\mathbf{B}\chi_b = b\chi_b$, then $\varphi_{ab} = \int \chi_b{}^* \psi_a dx = (\chi_b, \psi_a)$. Thus, φ_{ab} is an element $(a|b)$ of the transformation matrix for the transformation from a representation in which **A** is diagonal to one in which **B** is diagonal.

Probability can show the typical phenomena of interference, usually associated with waves, whose intensity is given by the square of the sum of contributions from different sources. The electron acts as a wave, (5), so to speak, as long as no attempt is made to verify that it is a particle; yet one can determine, if one wishes, by what route it travels just as though it were a particle; but when one does that, (4) applies and it does act like a particle.

These things are, of course, well known. They have already been explained many times.[9] However, it seems worth while to emphasize the fact that they are all simply direct consequences of Eq. (5), for it is essentially Eq. (5) that is fundamental in my formulation of quantum mechanics.

The generalization of Eqs. (4) and (5) to a large number of measurements, say $A, B, C, D, \cdots, K$, is, of course, that the probability of the sequence $a, b, c, d, \cdots, k$ is

$$P_{abcd\cdots k} = |\varphi_{abcd\cdots k}|^2.$$

The probability of the result a, c, k, for example, if $b, d, \cdots$ are measured, is the classical formula:

$$P_{ack} = \sum_b \sum_d \cdots P_{abcd\cdots k}, \quad (6)$$

while the probability of the same sequence a, c, k if no measurements are made between A and C and between C and K is

$$P_{ack}{}^q = |\sum_b \sum_d \cdots \varphi_{abcd\cdots k}|^2. \quad (7)$$

The quantity $\varphi_{abcd\cdots k}$ we can call the probability amplitude for the condition $A = a, B = b, C = c, D = d, \cdots, K = k$. (It is, of course, expressible as a product $\varphi_{ab}\varphi_{bc}\varphi_{cd}\cdots\varphi_{jk}$.)

3. THE PROBABILITY AMPLITUDE FOR A SPACE-TIME PATH

The physical ideas of the last section may be readily extended to define a probability amplitude for a particular completely specified space-time path. To explain how this may be done, we shall limit ourselves to a one-dimensional problem, as the generalization to several dimensions is obvious.

---

[9] See, for example, W. Heisenberg, *The Physical Principles of the Quantum Theory* (University of Chicago Press, Chicago, 1930), particularly Chapter IV.

Assume that we have a particle which can take up various values of a coordinate x. Imagine that we make an enormous number of successive position measurements, let us say separated by a small time interval ϵ. Then a succession of measurements such as $A, B, C, \cdots$ might be the succession of measurements of the coordinate x at successive times $t_1, t_2, t_3, \cdots$, where $t_{i+1} = t_i + \epsilon$. Let the value, which might result from measurement of the coordinate at time t_i, be x_i. Thus, if A is a measurement of x at t_1 then x_1 is what we previously denoted by a. From a classical point of view, the successive values, $x_1, x_2, x_3, \cdots$ of the coordinate practically define a path $x(t)$. Eventually, we expect to go the limit $\epsilon \to 0$.

The probability of such a path is a function of $x_1, x_2, \cdots, x_i, \cdots$, say $P(\cdots x_i, x_{i+1}, \cdots)$. The probability that the path lies in a particular region R of space-time is obtained classically by integrating P over that region. Thus, the probability that x_i lies between a_i and b_i, and x_{i+1} lies between a_{i+1} and b_{i+1}, etc., is

$$\cdots \int_{a_i}^{b_i} \int_{a_{i+1}}^{b_{i+1}} \cdots P(\cdots x_i, x_{i+1}, \cdots) \cdots dx_i dx_{i+1} \cdots$$

$$= \int_R P(\cdots x_i, x_{i+1}, \cdots) \cdots dx_i dx_{i+1} \cdots, \quad (8)$$

the symbol $\int_R$ meaning that the integration is to be taken over those ranges of the variables which lie within the region R. This is simply Eq. (6) with $a, b, \cdots$ replaced by $x_1, x_2, \cdots$ and integration replacing summation.

In quantum mechanics this is the correct formula for the case that $x_1, x_2, \cdots, x_i, \cdots$ were actually all measured, and then only those paths lying within R were taken. We would expect the result to be different if no such detailed measurements had been performed. Suppose a measurement is made which is capable only of determining that the path lies somewhere within R.

The measurement is to be what we might call an "ideal measurement." We suppose that no further details could be obtained from the same measurement without further disturbance to the system. I have not been able to find a precise definition. We are trying to avoid the extra uncertainties that must be averaged over if, for example, more information were measured but not utilized. We wish to use Eq. (5) or (7) for all x_i and have no residual part to sum over in the manner of Eq. (4).

We expect that the probability that the particle is found by our "ideal measurement" to be, indeed, in the region R is the square of a complex number $|\varphi(R)|^2$. The number $\varphi(R)$, which we may call the probability amplitude for region R is given by Eq. (7) with a, b, $\cdots$ replaced by x_i, x_{i+1}, $\cdots$ and summation replaced by integration:

$$\varphi(R) = \lim_{\epsilon \to 0} \int_R$$
$$\times \Phi(\cdots x_i, x_{i+1} \cdots) \cdots dx_i dx_{i+1} \cdots. \quad (9)$$

The complex number $\Phi(\cdots x_i, x_{i+1} \cdots)$ is a function of the variables x_i defining the path. Actually, we imagine that the time spacing ϵ approaches zero so that Φ essentially depends on the entire path $x(t)$ rather than only on just the values of x_i at the particular times t_i, $x_i = x(t_i)$. We might call Φ the probability amplitude functional of paths $x(t)$.

We may summarize these ideas in our first postulate:

I. If an ideal measurement is performed to determine whether a particle has a path lying in a region of space-time, then the probability that the result will be affirmative is the absolute square of a sum of complex contributions, one from each path in the region.

The statement of the postulate is incomplete. The meaning of a sum of terms one for "each" path is ambiguous. The precise meaning given in Eq. (9) is this: A path is first defined only by the positions x_i through which it goes at a sequence of equally spaced times,[10] $t_i = t_{i-1} + \epsilon$. Then all values of the coordinates within R have an equal weight. The actual magnitude of the weight depends upon ϵ and can be so chosen that the probability of an event which is certain

[10] There are very interesting mathematical problems involved in the attempt to avoid the subdivision and limiting processes. Some sort of complex measure is being associated with the space of functions $x(t)$. Finite results can be obtained under unexpected circumstances because the measure is not positive everywhere, but the contributions from most of the paths largely cancel out. These curious mathematical problems are sidestepped by the subdivision process. However, one feels as Cavalieri must have felt calculating the volume of a pyramid before the invention of calculus.

shall be normalized to unity. It may not be best to do so, but we have left this weight factor in a proportionality constant in the second postulate. The limit $\epsilon \to 0$ must be taken at the end of a calculation.

When the system has several degrees of freedom the coordinate space x has several dimensions so that the symbol x will represent a set of coordinates $(x^{(1)}, x^{(2)}, \cdots, x^{(k)})$ for a system with k degrees of freedom. A path is a sequence of configurations for successive times and is described by giving the configuration x_i or $(x_i^{(1)}, x_i^{(2)}, \cdots, x_i^{(k)})$, i.e., the value of each of the k coordinates for each time t_i. The symbol dx_i will be understood to mean the volume element in k dimensional configuration space (at time t_i). The statement of the postulates is independent of the coordinate system which is used.

The postulate is limited to defining the results of position measurements. It does not say what must be done to define the result of a momentum measurement, for example. This is not a real limitation, however, because in principle the measurement of momentum of one particle can be performed in terms of position measurements of other particles, e.g., meter indicators. Thus, an analysis of such an experiment will determine what it is about the first particle which determines its momentum.

4. THE CALCULATION OF THE PROBABILITY AMPLITUDE FOR A PATH

The first postulate prescribes the type of mathematical framework required by quantum mechanics for the calculation of probabilities. The second postulate gives a particular content to this framework by prescribing how to compute the important quantity Φ for each path:

II. The paths contribute equally in magnitude, but the phase of their contribution is the classical action (in units of $\hbar$); i.e., the time integral of the Lagrangian taken along the path.

That is to say, the contribution $\Phi[x(t)]$ from a given path $x(t)$ is proportional to $\exp(i/\hbar)S[x(t)]$, where the action $S[x(t)] = \int L(\dot{x}(t), x(t)) dt$ is the time integral of the classical Lagrangian $L(\dot{x}, x)$ taken along the path in question. The Lagrangian, which may be an explicit function of the time, is a function of position and velocity. If we suppose it to be a quadratic function of the velocities, we can show the mathematical equivalence of the postulates here and the more usual formulation of quantum mechanics.

To interpret the first postulate it was necessary to define a path by giving only the succession of points x_i through which the path passes at successive times t_i. To compute $S = \int L(\dot{x}, x)dt$ we need to know the path at all points, not just at x_i. We shall assume that the function $x(t)$ in the interval between t_i and t_{i+1} is the path followed by a classical particle, with the Lagrangian L, which starting from x_i at t_i reaches x_{i+1} at t_{i+1}. This assumption is required to interpret the second postulate for discontinuous paths. The quantity $\Phi(\cdots x_i, x_{i+1}, \cdots)$ can be normalized (for various ϵ) if desired, so that the probability of an event which is certain is normalized to unity as $\epsilon \to 0$.

There is no difficulty in carrying out the action integral because of the sudden changes of velocity encountered at the times t_i as long as L does not depend upon any higher time derivatives of the position than the first. Furthermore, unless L is restricted in this way the end points are not sufficient to define the classical path. Since the classical path is the one which makes the action a minimum, we can write

$$S = \sum_i S(x_{i+1}, x_i), \qquad (10)$$

where

$$S(x_{i+1}, x_i) = \mathrm{Min.} \int_{t_i}^{t_{i+1}} L(\dot{x}(t), x(t))dt. \qquad (11)$$

Written in this way, the only appeal to classical mechanics is to supply us with a Lagrangian function. Indeed, one could consider postulate two as simply saying, "Φ is the exponential of i times the integral of a real function of $x(t)$ and its first time derivative." Then the classical equations of motion might be derived later as the limit for large dimensions. The function of x and $\dot{x}$ then could be shown to be the classical Lagrangian within a constant factor.

Actually, the sum in (10), even for finite ϵ, is infinite and hence meaningless (because of the infinite extent of time). This reflects a further incompleteness of the postulates. We shall have to restrict ourselves to a finite, but arbitrarily long, time interval.

Combining the two postulates and using Eq. (10), we find

$$\varphi(R) = \underset{\epsilon \to 0}{\mathrm{Lim}} \int_R$$
$$\times \exp\left[\frac{i}{\hbar} \sum_i S(x_{i+1}, x_i)\right] \cdots \frac{dx_{i+1} \, dx_i}{A \quad A} \cdots, \qquad (12)$$

where we have let the normalization factor be split into a factor $1/A$ (whose exact value we shall presently determine) for each instant of time. The integration is just over those values $x_i, x_{i+1}, \cdots$ which lie in the region R. This equation, the definition (11) of $S(x_{i+1}, x_i)$, and the physical interpretation of $|\varphi(R)|^2$ as the probability that the particle will be found in R, complete our formulation of quantum mechanics.

5. DEFINITION OF THE WAVE FUNCTION

We now proceed to show the equivalence of these postulates to the ordinary formulation of quantum mechanics. This we do in two steps. We show in this section how the wave function may be defined from the new point of view. In the next section we shall show that this function satisfies Schroedinger's differential wave equation.

We shall see that it is the possibility, (10), of expressing S as a sum, and hence Φ as a product, of contributions from successive sections of the path, which leads to the possibility of defining a quantity having the properties of a wave function.

To make this clear, let us imagine that we choose a particular time t and divide the region R in Eq. (12) into pieces, future and past relative to t. We imagine that R can be split into: (a) a region R', restricted in any way in space, but lying entirely earlier in time than some t', such that $t' < t$; (b) a region R'' arbitrarily restricted in space but lying entirely later in time than t'', such that $t'' > t$; (c) the region between t' and t'' in which all the values of x coordinates are unrestricted, i.e., all of space-time between t' and t''. The region (c) is not absolutely necessary. It can be taken as narrow in time as desired. However, it is convenient in letting us consider varying t a little without having to redefine R' and R''. Then $|\varphi(R', R'')|^2$ is the probability that the path occupies R' and R''. Because R' is entirely previous to R'', considering the time t as the present, we can express this as the probability that the path had been in region R' and will be in region R''. If we divide by a factor, the probability that the path is in R', to renormalize the probability we find: $|\varphi(R', R'')|^2$ is the (relative) probability that if the system were in region R' it will be found later in R''.

This is, of course, the important quantity in predicting the results of many experiments. We prepare the system in a certain way (e.g., it was in region R') and then measure some other property (e.g., will it be found in region R''?). What does (12) say about computing this quantity, or rather the quantity $\varphi(R', R'')$ of which it is the square?

Let us suppose in Eq. (12) that the time t corresponds to one particular point k of the subdivision of time into steps ϵ, i.e., assume $t=t_k$, the index k, of course, depending upon the subdivision ϵ. Then, the exponential being the exponential of a sum may be split into a product of two factors

$$\exp\left[\frac{i}{\hbar}\sum_{i=k}^{\infty} S(x_{i+1}, x_i)\right]$$

$$\cdot \exp\left[\frac{i}{\hbar}\sum_{i=-\infty}^{k-1} S(x_{i+1}, x_i)\right]. \quad (13)$$

The first factor contains only coordinates with index k or higher, while the second contains only coordinates with index k or lower. This split is possible because of Eq. (10), which results essentially from the fact that the Lagrangian is a function only of positions and velocities. First, the integration on all variables x_i for $i > k$ can be performed on the first factor resulting in a function of x_k (times the second factor). Next, the integration on all variables x_i for $i < k$ can be performed on the second factor also, giving a function of x_k. Finally, the integration on x_k can be performed. That is, $\varphi(R', R'')$ can be written as the integral over x_k of the product of two factors. We will call these $\chi^*(x_k, t)$ and $\psi(x_k, t)$:

$$\varphi(R', R'') = \int \chi^*(x, t)\psi(x, t)dx, \quad (14)$$

where

$$\psi(x_k, t) = \lim_{\epsilon \to 0} \int_{R'}$$

$$\times \exp\left[\frac{i}{\hbar}\sum_{i=-\infty}^{k-1} S(x_{i+1}, x_i)\right]\frac{dx_{k-1}dx_{k-2}}{A}\cdots, \quad (15)$$

and

$$\chi^*(x_k, t) = \lim_{\epsilon \to 0} \int_{R''} \exp\left[\frac{i}{\hbar}\sum_{i=k}^{\infty} S(x_{i+1}, x_i)\right]$$

$$\cdot \frac{1}{A}\frac{dx_{k+1}dx_{k+2}}{A}\cdots. \quad (16)$$

The symbol R' is placed on the integral for ψ to indicate that the coordinates are integrated over the region R', and, for t_i between t' and t, over all space. In like manner, the integral for χ^* is over R'' and over all space for those coordinates corresponding to times between t and t''. The asterisk on χ^* denotes complex conjugate, as it will be found more convenient to define (16) as the complex conjugate of some quantity, χ.

The quantity ψ depends only upon the region R' previous to t, and is completely defined if that region is known. It does not depend, in any way, upon what will be done to the system after time t. This latter information is contained in χ. Thus, with ψ and χ we have separated the past history from the future experiences of the system. This permits us to speak of the relation of past and future in the conventional manner. Thus, if a particle has been in a region of space-time R' it may at time t be said to be in a certain condition, or state, determined only by its past and described by the so-called wave function $\psi(x, t)$. This function contains all that is needed to predict future probabilities. For, suppose, in another situation, the region R' were different, say r', and possibly the Lagrangian for times before t were also altered. But, nevertheless, suppose the quantity from Eq. (15) turned out to be the same. Then, according to (14) the probability of ending in any region R'' is the same for R' as for r'. Therefore, future measurements will not distinguish whether the system had occupied R' or r'. Thus, the wave function $\psi(x, t)$ is sufficient to define those attributes which are left from past history which determine future behavior.

Likewise, the function $\chi^*(x, t)$ characterizes the experience, or, let us say, experiment to which the system is to be subjected. If a different region, r'' and different Lagrangian after t, were to give the same $\chi^*(x, t)$ via Eq. (16), as does region R'', then no matter what the preparation, ψ, Eq. (14) says that the chance of finding the system in R'' is always the same as finding it in r''. The two "experiments" R'' and r'' are equivalent, as they yield the same results. We shall say loosely that these experiments are to determine with what probability the system is in state χ. Actually, this terminology is poor. The system is really in state ψ. The reason we can associate a state with an experiment is, of course, that for an ideal experiment there turns out to be a unique state (whose wave function is $\chi(x, t)$) for which the experiment succeeds with certainty.

Thus, we can say: the probability that a system in state ψ will be found by an experiment whose characteristic state is χ (or, more loosely, the chance that a system in state ψ will appear to be in χ) is

$$\left| \int \chi^*(x, t)\psi(x, t)dx \right|^2. \quad (17)$$

These results agree, of course, with the principles of ordinary quantum mechanics. They are a consequence of the fact that the Lagrangian is a function of position, velocity, and time only.

6. THE WAVE EQUATION

To complete the proof of the equivalence with the ordinary formulation we shall have to show that the wave function defined in the previous section by Eq. (15) actually satisfies the Schroedinger wave equation. Actually, we shall only succeed in doing this when the Lagrangian L in (11) is a quadratic, but perhaps inhomogeneous, form in the velocities $\dot{x}(t)$. This is not a limitation, however, as it includes all the cases for which the Schroedinger equation has been verified by experiment.

The wave equation describes the development of the wave function with time. We may expect to approach it by noting that, for finite ϵ, Eq. (15) permits a simple recursive relation to be developed. Consider the appearance of Eq. (15) if we were to compute ψ at the next instant of time:

$$\psi(x_{k+1}, t+\epsilon) = \int_{R^1} \exp\left[\frac{i}{\hbar} \sum_{i=-\infty}^{k} S(x_{i+1}, x_i)\right]$$

$$\times \frac{dx_k}{A} \frac{dx_{k-1}}{A} \cdots . \quad (15')$$

This is similar to (15) except for the integration over the additional variable x_k and the extra term in the sum in the exponent. This term means that the integral of (15') is the same as the integral of (15) except for the factor $(1/A) \exp(i/\hbar)S(x_{k+1}, x_k)$. Since this does not contain any of the variables x_i for i less than k, all of the integrations on dx_i up to dx_{k-1} can be performed with this factor left out. However, the result of these integrations is by (15) simply $\psi(x_k, t)$. Hence, we find from (15') the relation

$$\psi(x_{k+1}, t+\epsilon)$$

$$= \int \exp\left[\frac{i}{\hbar}S(x_{k+1}, x_k)\right]\psi(x_k, t)dx_k/A. \quad (18)$$

This relation giving the development of ψ with time will be shown, for simple examples, with suitable choice of A, to be equivalent to Schroedinger's equation. Actually, Eq. (18) is not exact, but is only true in the limit $\epsilon \to 0$ and we shall derive the Schroedinger equation by assuming (18) is valid to first order in ϵ. The Eq. (18) need only be true for small ϵ to the first order in ϵ. For if we consider the factors in (15) which carry us over a finite interval of time, T, the number of factors is T/ϵ. If an error of order ϵ^2 is made in each, the resulting error will not accumulate beyond the order $\epsilon^2(T/\epsilon)$ or $T\epsilon$, which vanishes in the limit.

We shall illustrate the relation of (18) to Schroedinger's equation by applying it to the simple case of a particle moving in one dimension in a potential $V(x)$. Before we do this, however, we would like to discuss some approximations to the value $S(x_{i+1}, x_i)$ given in (11) which will be sufficient for expression (18).

The expression defined in (11) for $S(x_{i+1}, x_i)$ is difficult to calculate exactly for arbitrary ϵ from classical mechanics. Actually, it is only necessary that an approximate expression for $S(x_{i+1}, x_i)$ be used in (18), provided the error of the approximation be of an order smaller than the first in ϵ. We limit ourselves to the case that the Lagrangian is a quadratic, but perhaps inhomogeneous, form in the velocities $\dot{x}(t)$. As we shall see later, the paths which are important are those for which $x_{i+1}-x_i$ is of order $\epsilon^{\frac{1}{2}}$. Under these circumstances, it is sufficient to calculate the integral in (11) over the classical path taken by a *free* particle.[11]

In *Cartesian coordinates*[12] the path of a free particle is a straight line so the integral of (11) can be taken along a straight line. Under these circumstances it is sufficiently accurate to replace the integral by the trapezoidal rule

$$S(x_{i+1}, x_i) = \frac{\epsilon}{2} L\left(\frac{x_{i+1}-x_i}{\epsilon}, x_{i+1}\right)$$
$$+ \frac{\epsilon}{2} L\left(\frac{x_{i+1}-x_i}{\epsilon}, x_i\right) \quad (19)$$

or, if it proves more convenient,

$$S(x_{i+1}, x_i) = \epsilon L\left(\frac{x_{i+1}-x_i}{\epsilon}, \frac{x_{i+1}+x_i}{2}\right). \quad (20)$$

These are not valid in a general coordinate system, e.g., spherical. An even simpler approximation may be used if, in addition, there is no vector potential or other terms linear in the velocity (see page 376):

$$S(x_{i+1}, x_i) = \epsilon L\left(\frac{x_{i+1}-x_i}{\epsilon}, x_{i+1}\right). \quad (21)$$

Thus, for the simple example of a particle of mass m moving in one dimension under a potential $V(x)$, we can set

$$S(x_{i+1}, x_i) = \frac{m\epsilon}{2}\left(\frac{x_{i+1}-x_i}{\epsilon}\right)^2 - \epsilon V(x_{i+1}). \quad (22)$$

---

[11] It is assumed that the "forces" enter through a scalar and vector potential and not in terms involving the square of the velocity. More generally, what is meant by a free particle is one for which the Lagrangian is altered by omission of the terms linear in, and those independent of, the velocities.

[12] More generally, coordinates for which the terms quadratic in the velocity in $L(\dot{x}, x)$ appear with constant coefficients.

For this example, then, Eq. (18) becomes

$$\psi(x_{k+1}, t+\epsilon) = \int \exp\left[\frac{i\epsilon}{\hbar}\left\{\frac{m}{2}\left(\frac{x_{k+1}-x_k}{\epsilon}\right)^2 - V(x_{k+1})\right\}\right]\psi(x_k, t)dx_k/A. \quad (23)$$

Let us call $x_{k+1}=x$ and $x_{k+1}-x_k=\xi$ so that $x_k = x-\xi$. Then (23) becomes

$$\psi(x, t+\epsilon) = \int \exp\frac{im\xi^2}{\epsilon \cdot 2\hbar}$$
$$\cdot \exp\frac{-i\epsilon V(x)}{\hbar} \cdot \psi(x-\xi, t)\frac{d\xi}{A}. \quad (24)$$

The integral on ξ will converge if $\psi(x, t)$ falls off sufficiently for large x (certainly if $\int \psi^*(x)\psi(x)dx = 1$). In the integration on ξ, since ϵ is very small, the exponential of $im\xi^2/2\hbar\epsilon$ oscillates extremely rapidly except in the region about $\xi=0$ (ξ of order $(\hbar\epsilon/m)^{\frac{1}{2}}$). Since the function $\psi(x-\xi, t)$ is a relatively smooth function of ξ (since ϵ may be taken as small as desired), the region where the exponential oscillates rapidly will contribute very little because of the almost complete cancelation of positive and negative contributions. Since only small ξ are effective, $\psi(x-\xi, t)$ may be expanded as a Taylor series. Hence,

$$\psi(x, t+\epsilon) = \exp\left(\frac{-i\epsilon V(x)}{\hbar}\right)$$
$$\times \int \exp\left(\frac{im\xi^2}{2\hbar\epsilon}\right)\left[\psi(x, t) - \xi\frac{\partial\psi(x, t)}{\partial x}\right.$$
$$\left. + \frac{\xi^2}{2}\frac{\partial^2\psi(x, t)}{\partial x^2} - \cdots\right]d\xi/A. \quad (25)$$

Now

$$\int_{-\infty}^{\infty} \exp(im\xi^2/2\hbar\epsilon)d\xi = (2\pi\hbar\epsilon i/m)^{\frac{1}{2}},$$

$$\int_{-\infty}^{\infty} \exp(im\xi^2/2\hbar\epsilon)\xi d\xi = 0, \quad (26)$$

$$\int_{-\infty}^{\infty} \exp(im\xi^2/2\hbar\epsilon)\xi^2 d\xi = (\hbar\epsilon i/m)(2\pi\hbar\epsilon i/m)^{\frac{1}{2}},$$

while the integral containing ξ^3 is zero, for like

the one with ξ it possesses an odd integrand, and the ones with ξ^4 are of at least the order ϵ smaller than the ones kept here.[13] If we expand the left-hand side to first order in ϵ, (25) becomes

$$\psi(x,t)+\epsilon\frac{\partial\psi(x,t)}{\partial t}$$
$$=\exp\left(\frac{-i\epsilon V(x)}{\hbar}\right)\frac{(2\pi\hbar \epsilon i/m)^{\frac{1}{2}}}{A}$$
$$\times\left[\psi(x,t)+\frac{\hbar \epsilon i}{2m}\frac{\partial^2\psi(x,t)}{\partial x^2}+\cdots\right]. \quad (27)$$

In order that both sides may agree to *zero* order in ϵ, we must set

$$A=(2\pi\hbar \epsilon i/m)^{\frac{1}{2}}. \quad (28)$$

Then expanding the exponential containing $V(x)$, we get

$$\psi(x,t)+\epsilon\frac{\partial\psi}{\partial t}=\left(1-\frac{i\epsilon}{\hbar}V(x)\right)$$
$$\times\left(\psi(x,t)+\frac{\hbar \epsilon i}{2m}\frac{\partial^2\psi}{\partial x^2}\right). \quad (29)$$

Canceling $\psi(x,t)$ from both sides, and comparing terms to first order in ϵ and multiplying by $-\hbar/i$ one obtains

$$-\frac{\hbar}{i}\frac{\partial\psi}{\partial t}=\frac{1}{2m}\left(\frac{\hbar}{i}\frac{\partial}{\partial x}\right)^2\psi+V(x)\psi, \quad (30)$$

which is Schroedinger's equation for the problem in question.

The equation for χ^* can be developed in the same way, but adding a factor *decreases* the time by one step, i.e., χ^* satisfies an equation like (30) but with the sign of the time reversed. By taking complex conjugates we can conclude that χ satisfies the same equation as ψ, i.e., an experiment can be defined by the particular state χ to which it corresponds.[14]

This example shows that most of the contribution to $\psi(x_{k+1}, t+\epsilon)$ comes from values of x_k in $\psi(x_k, t)$ which are quite close to x_{k+1} (distant of order $\epsilon^{\frac{1}{2}}$) so that the integral equation (23) can, in the limit, be replaced by a differential equation. The "velocities," $(x_{k+1}-x_k)/\epsilon$ which are important are very high, being of order $(\hbar/m\epsilon)^{\frac{1}{2}}$ which diverges as $\epsilon\to 0$. The paths involved are, therefore, continuous but possess no derivative. They are of a type familiar from study of Brownian motion.

It is these large velocities which make it so necessary to be careful in approximating $S(x_{k+1}, x_k)$ from Eq. (11).[15] To replace $V(x_{k+1})$ by $V(x_k)$ would, of course, change the exponent in (18) by $i\epsilon[V(x_k)-V(x_{k+1})]/\hbar$ which is of order $\epsilon(x_{k+1}-x_k)$, and thus lead to unimportant terms of higher order than ϵ on the right-hand side of (29). It is for this reason that (20) and (21) are equally satisfactory approximations to $S(x_{i+1}, x_i)$ when there is no vector potential. A term, linear in velocity, however, arising from a vector potential, as $A\dot{x}dt$ must be handled more carefully. Here a term in $S(x_{k+1}, x_k)$ such as $A(x_{k+1})\times(x_{k+1}-x_k)$ differs from $A(x_k)(x_{k+1}-x_k)$ by a term of order $(x_{k+1}-x_k)^2$, and, therefore, of order ϵ. Such a term would lead to a change in the resulting wave equation. For this reason the approximation (21) is not a sufficiently accurate approximation to (11) and one like (20), (or (19) from which (20) differs by terms of order higher than ϵ) must be used. If **A** represents the vector potential and $\mathbf{p}=(\hbar/i)\nabla$, the momentum operator, then (20) gives, in the Hamiltonian operator, a term $(1/2m)(\mathbf{p}-(e/c)\mathbf{A})\cdot(\mathbf{p}-(e/c)\mathbf{A})$, while (21) gives $(1/2m)(\mathbf{p}\cdot\mathbf{p}-(2e/c)\mathbf{A}\cdot\mathbf{p}+(e^2/c^2)\mathbf{A}\cdot\mathbf{A})$. These two expressions differ by $(\hbar e/2imc)\nabla\cdot\mathbf{A}$

---

[13] Really, these integrals are oscillatory and not defined, but they may be defined by using a convergence factor. Such a factor is automatically provided by $\psi(x-\xi, t)$ in (24). If a more formal procedure is desired replace $\hbar$ by $\hbar(1-i\delta)$, for example, where δ is a small positive number, and then let $\delta\to 0$.

[14] Dr. Hartland Snyder has pointed out to me, in private conversation, the very interesting possibility that there may be a generalization of quantum mechanics in which the states measured by experiment cannot be prepared; that is, there would be no state into which a system may be put for which a particular experiment gives certainty for a result. The class of functions χ is not identical to the class of available states ψ. This would result if, for example, χ satisfied a different equation than ψ.

[15] Equation (18) is actually exact when (11) is used for $S(x_{i+1}, x_i)$ for arbitrary ϵ for cases in which the potential does not involve x to higher powers than the second (e.g., free particle, harmonic oscillator). It is necessary, however, to use a more accurate value of A. One can define A in this way. Assume classical particles with k degrees of freedom start from the point x_i, t_i with uniform density in momentum space. Write the number of particles having a given component of momentum in range dp as dp/p_0 with p_0 constant. Then $A=(2\pi\hbar i/p_0)^{k/2}\rho^{-\frac{1}{2}}$, where ρ is the density in k dimensional coordinate space x_{i+1} of these particles at time t_{i+1}.

which may not be zero. The question is still more important in the coefficient of terms which are quadratic in the velocities. In these terms (19) and (20) are not sufficiently accurate representations of (11) in general. It is when the coefficients are constant that (19) or (20) can be substituted for (11). If an expression such as (19) is used, say for spherical coordinates, when it is not a valid approximation to (11), one obtains a Schroedinger equation in which the Hamiltonian operator has some of the momentum operators and coordinates in the wrong order. Equation (11) then resolves the ambiguity in the usual rule to replace p and q by the non-commuting quantities $(\hbar/i)(\partial/\partial q)$ and q in the classical Hamiltonian $H(p, q)$.

It is clear that the statement (11) is independent of the coordinate system. Therefore, to find the differential wave equation it gives in any coordinate system, the easiest procedure is first to find the equations in Cartesian coordinates and then to transform the coordinate system to the one desired. It suffices, therefore, to show the relation of the postulates and Schroedinger's equation in rectangular coordinates.

The derivation given here for one dimension can be extended directly to the case of three-dimensional Cartesian coordinates for any number, K, of particles interacting through potentials with one another, and in a magnetic field, described by a vector potential. The terms in the vector potential require completing the square in the exponent in the usual way for Gaussian integrals. The variable x must be replaced by the set $x^{(1)}$ to $x^{(3K)}$ where $x^{(1)}$, $x^{(2)}$, $x^{(3)}$ are the coordinates of the first particle of mass m_1, $x^{(4)}$, $x^{(5)}$, $x^{(6)}$ of the second of mass m_2, etc. The symbol dx is replaced by $dx^{(1)}dx^{(2)}\cdots dx^{(3K)}$, and the integration over dx is replaced by a $3K$-fold integral. The constant A has, in this case, the value $A = (2\pi\hbar\epsilon i/m_1)^{\frac{1}{2}}(2\pi\hbar\epsilon i/m_2)^{\frac{3}{2}}\cdots(2\pi\hbar\epsilon i/m_K)^{\frac{3}{2}}$. The Lagrangian is the classical Lagrangian for the same problem, and the Schroedinger equation resulting will be that which corresponds to the classical Hamiltonian, derived from this Lagrangian. The equations in any other coordinate system may be obtained by transformation. Since this includes all cases for which Schroedinger's equation has been checked with experiment, we may say our postulates are able to describe what can be described by non-relativistic quantum mechanics, neglecting spin.

7. DISCUSSION OF THE WAVE EQUATION

The Classical Limit

This completes the demonstration of the equivalence of the new and old formulations. We should like to include in this section a few remarks about the important equation (18).

This equation gives the development of the wave function during a small time interval. It is easily interpreted physically as the expression of Huygens' principle for matter waves. In geometrical optics the rays in an inhomogeneous medium satisfy Fermat's principle of least *time*. We may state Huygens' principle in wave optics in this way: If the amplitude of the wave is known on a given surface, the amplitude at a near by point can be considered as a sum of contributions from all points of the surface. Each contribution is delayed in phase by an amount proportional to the *time* it would take the light to get from the surface to the point along the ray of least *time* of geometrical optics. We can consider (22) in an analogous manner starting with Hamilton's first principle of least *action* for classical or "geometrical" mechanics. If the amplitude of the wave ψ is known on a given "surface," in particular the "surface" consisting of all x at time t, its value at a particular nearby point at time $t+\epsilon$, is a sum of contributions from all points of the surface at t. Each contribution is delayed in phase by an amount proportional to the *action* it would require to get from the surface to the point along the path of least *action* of classical mechanics.[16]

Actually Huygens' principle is not correct in optics. It is replaced by Kirchoff's modification which requires that both the amplitude and its derivative must be known on the adjacent surface. This is a consequence of the fact that the wave equation in optics is second order in the time. The wave equation of quantum mechanics is first order in the time; therefore, Huygens' principle *is* correct for matter waves, action replacing time.

---
[16] See in this connection the very interesting remarks of Schroedinger, Ann. d. Physik 79, 489 (1926).

The equation can also be compared mathematically to quantities appearing in the usual formulations. In Schroedinger's method the development of the wave function with time is given by

$$-\frac{\hbar}{i}\frac{\partial \psi}{\partial t} = \mathbf{H}\psi, \qquad (31)$$

which has the solution (for any ϵ if $\mathbf{H}$ is time independent)

$$\psi(x, t+\epsilon) = \exp(-i\epsilon\mathbf{H}/\hbar)\psi(x, t). \qquad (32)$$

Therefore, Eq. (18) expresses the operator $\exp(-i\epsilon\mathbf{H}/\hbar)$ by an approximate integral operator for small ϵ.

From the point of view of Heisenberg one considers the position at time t, for example, as an operator $\mathbf{x}$. The position $\mathbf{x}'$ at a later time $t+\epsilon$ can be expressed in terms of that at time t by the operator equation

$$\mathbf{x}' = \exp(i\epsilon\mathbf{H}/\hbar)\mathbf{x}\exp-(i\epsilon\mathbf{H}/\hbar). \qquad (33)$$

The transformation theory of Dirac allows us to consider the wave function at time $t+\epsilon$, $\psi(x', t+\epsilon)$, as representing a state in a representation in which $\mathbf{x}'$ is diagonal, while $\psi(x, t)$ represents the same state in a representation in which $\mathbf{x}$ is diagonal. They are, therefore, related through the transformation function $(x'|x)_\epsilon$ which relates these representations:

$$\psi(x', t+\epsilon) = \int (x'|x)_\epsilon \psi(x,t)\,dx.$$

Therefore, the content of Eq. (18) is to show that for small ϵ we can set

$$(x'|x)_\epsilon = (1/A)\exp(iS(x', x)/\hbar) \qquad (34)$$

with $S(x', x)$ defined as in (11).

The close analogy between $(x'|x)_\epsilon$ and the quantity $\exp(iS(x', x)/\hbar)$ has been pointed out on several occasions by Dirac.[1] In fact, we now see that to sufficient approximations the two quantities may be taken to be proportional to each other. Dirac's remarks were the starting point of the present development. The points he makes concerning the passage to the classical limit $\hbar \to 0$ are very beautiful, and I may perhaps be excused for briefly reviewing them here.

First we note that the wave function at x'' at time t'' can be obtained from that at x' at time t' by

$$\psi(x'', t'') = \underset{\epsilon \to 0}{\text{Lim}} \int \cdots \int$$

$$\times \exp\left[\frac{i}{\hbar}\sum_{i=0}^{j-1} S(x_{i+1}, x_i)\right]$$

$$\times \psi(x', t')\frac{dx_0}{A}\frac{dx_1}{A}\cdots\frac{dx_{j-1}}{A}, \qquad (35)$$

where we put $x_0 \equiv x'$ and $x_j \equiv x''$ where $j\epsilon = t'' - t'$ (between the times t' and t'' we assume no restriction is being put on the region of integration). This can be seen either by repeated applications of (18) or directly from Eq. (15). Now we ask, as $\hbar \to 0$ what values of the intermediate coordinates x_i contribute most strongly to the integral? These will be the values most likely to be found by experiment and therefore will determine, in the limit, the classical path. If $\hbar$ is very small, the exponent will be a very rapidly varying function of any of its variables x_i. As x_i varies, the positive and negative contributions of the exponent nearly cancel. The region at which x_i contributes most strongly is that at which the phase of the exponent varies least rapidly with x_i (method of stationary phase). Call the sum in the exponent S;

$$S = \sum_{i=0}^{j-1} S(x_{i+1}, x_i). \qquad (36)$$

Then the classical orbit passes, approximately, through those points x_i at which the rate of change of S with x_i is small, or in the limit of small $\hbar$, zero, i.e., the classical orbit passes through the points at which $\partial S/\partial x_i = 0$ for all x_i. Taking the limit $\epsilon \to 0$, (36) becomes in view of (11)

$$S = \int_{t'}^{t''} L(\dot{x}(t), x(t))dt. \qquad (37)$$

We see then that the classical path is that for which the integral (37) suffers no first-order change on varying the path. This is Hamilton's principle and leads directly to the Lagrangian equations of motion.

8. OPERATOR ALGEBRA

Matrix Elements

Given the wave function and Schroedinger's equation, of course all of the machinery of operator or matrix algebra can be developed. It is, however, rather interesting to express these concepts in a somewhat different language more closely related to that used in stating the postulates. Little will be gained by this in elucidating operator algebra. In fact, the results are simply a translation of simple operator equations into a somewhat more cumbersome notation. On the other hand, the new notation and point of view are very useful in certain applications described in the introduction. Furthermore, the form of the equations permits natural extension to a wider class of operators than is usually considered (e.g., ones involving quantities referring to two or more different times). If any generalization to a wider class of action functionals is possible, the formulae to be developed will play an important role.

We discuss these points in the next three sections. This section is concerned mainly with definitions. We shall define a quantity which we call a transition element between two states. It is essentially a matrix element. But instead of being the matrix element between a state ψ and another χ corresponding to the *same* time, these two states will refer to different times. In the following section a fundamental relation between transition elements will be developed from which the usual commutation rules between coordinate and momentum may be deduced. The same relation also yields Newton's equation of motion in matrix form. Finally, in Section 10 we discuss the relation of the Hamiltonian to the operation of displacement in time.

We begin by defining a transition element in terms of the probability of transition from one state to another. More precisely, suppose we have a situation similar to that described in deriving (17). The region R consists of a region R' previous to t', all space between t' and t'' and the region R'' after t''. We shall study the probability that a system in region R' is later found in region R''. This is given by (17). We shall discuss in this section how it changes with changes in the form of the Lagrangian between t' and t''. In Section 10 we discuss how it changes with changes in the preparation R' or the experiment R''.

The state at time t' is defined completely by the preparation R'. It can be specified by a wave function $\psi(x', t')$ obtained as in (15), but containing only integrals up to the time t'. Likewise, the state characteristic of the experiment (region R'') can be defined by a function $\chi(x'', t'')$ obtained from (16) with integrals only beyond t''. The wave function $\psi(x'', t'')$ at time t'' can, of course, also be gotten by appropriate use of (15). It can also be gotten from $\psi(x', t')$ by (35). According to (17) with t'' used instead of t, the probability of being found in χ if prepared in ψ is the square of what we shall call the transition amplitude $\int \chi^*(x'', t'') \psi(x'', t'') dx''$. We wish to express this in terms of χ at t'' and ψ at t'. This we can do with the aid of (35). Thus, the chance that a system prepared in state $\psi_{t'}$ at time t' will be found after t'' to be in a state $\chi_{t''}$ is the square of the transition amplitude

$$\langle \chi_{t''} | 1 | \psi_{t'} \rangle_S = \lim_{\epsilon \to 0} \int \cdots \int \chi^*(x'', t'')$$
$$\times \exp(iS/\hbar) \psi(x', t') \frac{dx_0}{A} \cdots \frac{dx_{j-1}}{A} dx_j, \quad (38)$$

where we have used the abbreviation (36).

In the language of ordinary quantum mechanics if the Hamiltonian, $\mathbf{H}$, is constant, $\psi(x, t'') = \exp[-i(t''-t')\mathbf{H}/\hbar]\psi(x, t')$ so that (38) is the matrix element of $\exp[-i(t''-t')\mathbf{H}/\hbar]$ between states $\chi_{t''}$ and $\psi_{t'}$.

If F is any function of the coordinates x_i for $t' < t_i < t''$, we shall define the transition element of F between the states ψ at t' and χ at t'' for the action S as $(x'' \equiv x_j, x' \equiv x_0)$:

$$\langle \chi_{t''} | F | \psi_{t'} \rangle_S = \lim_{\epsilon \to 0} \int \cdots \int$$
$$\times \chi^*(x'', t'') F(x_0, x_1, \cdots x_j)$$
$$\cdot \exp\left[\frac{i}{\hbar}\sum_{i=0}^{j-1} S(x_{i+1}, x_i)\right] \psi(x', t') \frac{dx_0}{A} \cdots \frac{dx_{j-1}}{A} dx_j. \quad (39)$$

In the limit $\epsilon \to 0$, F is a functional of the path $x(t)$.

We shall see presently why such quantities are important. It will be easier to understand if we stop for a moment to find out what the quantities correspond to in conventional notation. Suppose F is simply x_k where k corresponds to some time $t=t_k$. Then on the right-hand side of (39) the integrals from x_0 to x_{k-1} may be performed to produce $\psi(x_k, t)$ or $\exp[-i(t-t')\mathbf{H}/\hbar]\psi_{t'}$. In like manner the integrals on x_i for $j \geqslant i > k$ give $\chi^*(x_k, t)$ or $\{\exp[-i(t''-t)\mathbf{H}/\hbar]\chi_{t''}\}^*$. Thus, the transition element of x_k,

$$\langle \chi_{t''}|F|\psi_{t'}\rangle_S$$

$$= \int \chi_{t''}{}^* e^{-(i/\hbar)\mathbf{H}(t''-t)} x e^{-(i/\hbar)\mathbf{H}(t-t')} \psi_{t'} dx$$

$$= \int \chi^*(x, t) x \psi(x, t) dx \quad (40)$$

is the matrix element of $\mathbf{x}$ at time $t = t_k$ between the state which would develop at time t from $\psi_{t'}$ at t' and the state which will develop from time t to $\chi_{t''}$ at t''. It is, therefore, the matrix element of $\mathbf{x}(t)$ between these states.

Likewise, according to (39) with $F = x_{k+1}$, the transition element of x_{k+1} is the matrix element of $\mathbf{x}(t+\epsilon)$. The transition element of $F = (x_{k+1} - x_k)/\epsilon$ is the matrix element of $(\mathbf{x}(t+\epsilon) - \mathbf{x}(t))/\epsilon$ or of $i(\mathbf{Hx} - \mathbf{xH})/\hbar$, as is easily shown from (40). We can call this the matrix element of velocity $\dot{x}(t)$.

Suppose we consider a second problem which differs from the first because, for example, the potential is augmented by a small amount $U(, \mathbf{x}t)$. Then in the new problem the quantity replacing S is $S' = S + \sum_i \epsilon U(x_i, t_i)$. Substitution into (38) leads directly to

$$\langle \chi_{t''}|1|\psi_{t'}\rangle_{S'}$$

$$= \left\langle \chi_{t''} \middle| \exp\frac{i\epsilon}{\hbar} \sum_{i=1}^{j} U(x_i, t_i) \middle| \psi_{t'} \right\rangle_S. \quad (41)$$

Thus, transition elements such as (39) are important insofar as F may arise in some way from a change δS in an action expression. We denote, by observable functionals, those functionals F which can be defined, (possibly indirectly) in terms of the changes which are produced by possible changes in the action S. The condition that a functional be observable is somewhat similar to the condition that an operator be Hermitian. The observable functionals are a restricted class because the action must remain a quadratic function of velocities. From one observable functional others may be derived, for example, by

$$\langle \chi_{t''}|F|\psi_{t'}\rangle_{S'}$$

$$= \left\langle \chi_{t''} \middle| F \exp\frac{i\epsilon}{\hbar} \sum_{i=1}^{j} U(x_i, t_i) \middle| \psi_{t'} \right\rangle_S \quad (42)$$

which is obtained from (39).

Incidentally, (41) leads directly to an important perturbation formula. If the effect of U is small the exponential can be expanded to first order in U and we find

$$\langle \chi_{t''}|1|\psi_{t'}\rangle_{S'} = \langle \chi_{t''}|1|\psi_{t'}\rangle_S$$

$$+ \frac{i}{\hbar}\langle \chi_{t''}|\sum_i \epsilon U(x_i, t_i)|\psi_{t'}\rangle. \quad (43)$$

Of particular importance is the case that $\chi_{t''}$ is a state in which $\psi_{t'}$ would not be found at all were it not for the disturbance, U (i.e., $\langle \chi_{t''}|1|\psi_{t'}\rangle_S = 0$). Then

$$\frac{1}{\hbar^2}|\langle \chi_{t''}|\sum_i \epsilon U(x_i, t_i)|\psi_{t'}\rangle_S|^2 \quad (44)$$

is the probability of transition as induced to first order by the perturbation. In ordinary notation,

$$\langle \chi_{t''}|\sum_i \epsilon U(x_i, t_i)|\psi_{t'}\rangle_S$$

$$= \int \left\{ \int \chi_{t''}{}^* e^{-(i/\hbar)\mathbf{H}(t''-t)} \mathbf{U} e^{-(i/\hbar)\mathbf{H}(t-t')} \psi_{t'} dx \right\} dt$$

so that (44) reduces to the usual expression[17] for time dependent perturbations.

9. NEWTON'S EQUATIONS
The Commutation Relation

In this section we find that different functionals may give identical results when taken between any two states. This equivalence between functionals is the statement of operator equations in the new language.

If F depends on the various coordinates, we can, of course, define a new functional $\partial F/\partial x_k$

---

[17] P. A. M. Dirac, *The Principles of Quantum Mechanics* (The Clarendon Press, Oxford, 1935), second edition, Section 47, Eq. (20).

by differentiating it with respect to one of its variables, say x_k $(0<k<j)$. If we calculate $\langle \chi_{t''} | \partial F/\partial x_k | \psi_{t'} \rangle_S$ by (39) the integral on the right-hand side will contain $\partial F/\partial x_k$. The only other place that the variable x_k appears is in S. Thus, the integration on x_k can be performed by parts. The integrated part vanishes (assuming wave functions vanish at infinity) and we are left with the quantity $-F(\partial/\partial x_k)\exp(iS/\hbar)$ in the integral. However, $(\partial/\partial x_k)\exp(iS/\hbar) = (i/\hbar)(\partial S/\partial x_k)\exp(iS/\hbar)$, so the right side represents the transition element of $-(i/\hbar)F(\partial S/\partial x_k)$, i.e.,

$$\left\langle \chi_{t''} \left| \frac{\partial F}{\partial x_k} \right| \psi_{t'} \right\rangle_S = -\frac{i}{\hbar} \left\langle \chi_{t''} \left| F \frac{\partial S}{\partial x_k} \right| \psi_{t'} \right\rangle_S. \quad (45)$$

This very important relation shows that two different functionals may give the same result for the transition element between any two states. We say they are equivalent and symbolize the relation by

$$-\frac{\hbar}{i}\frac{\partial F}{\partial x_k} \underset{S}{\leftrightarrow} F\frac{\partial S}{\partial x_k}, \quad (46)$$

the symbol $\underset{S}{\leftrightarrow}$ emphasizing the fact that functionals equivalent under one action may not be equivalent under another. The quantities in (46) need not be observable. The equivalence is, nevertheless, true. Making use of (36) one can write

$$-\frac{\hbar}{i}\frac{\partial F}{\partial x_k} \underset{S}{\leftrightarrow} F\left[\frac{\partial S(x_{k+1}, x_k)}{\partial x_k} + \frac{\partial S(x_k, x_{k-1})}{\partial x_k}\right]. \quad (47)$$

This equation is true to zero and first order in ϵ and has as consequences the commutation relations of momentum and coordinate, as well as the Newtonian equations of motion in matrix form.

In the case of our simple one-dimensional problem, $S(x_{i+1}, x_i)$ is given by the expression (15), so that

$$\partial S(x_{k+1}, x_k)/\partial x_k = -m(x_{k+1}-x_k)/\epsilon,$$

and

$$\partial S(x_k, x_{k-1})/\partial x_k = +m(x_k-x_{k-1})/\epsilon - \epsilon V'(x_k);$$

where we write $V'(x)$ for the derivative of the potential, or force. Then (47) becomes

$$-\frac{\hbar}{i}\frac{\partial F}{\partial x_k} \underset{S}{\leftrightarrow} F\left[-m\left(\frac{x_{k+1}-x_k}{\epsilon}\right.\right.$$
$$\left.\left. -\frac{x_k-x_{k-1}}{\epsilon}\right) - \epsilon V'(x_k)\right]. \quad (48)$$

If F does not depend on the variable x_k, this gives Newton's equations of motion. For example, if F is constant, say unity, (48) just gives (dividing by ϵ)

$$0 \underset{S}{\leftrightarrow} -\frac{m}{\epsilon}\left(\frac{x_{k+1}-x_k}{\epsilon} - \frac{x_k-x_{k-1}}{\epsilon}\right) - V'(x_k).$$

Thus, the transition element of mass times acceleration $[(x_{k+1}-x_k)/\epsilon - (x_k-x_{k-1})/\epsilon]/\epsilon$ between any two states is equal to the transition element of force $-V'(x_k)$ between the same states. This is the matrix expression of Newton's law which holds in quantum mechanics.

What happens if F does depend upon x_k? For example, let $F = x_k$. Then (48) gives, since $\partial F/\partial x_k = 1$,

$$-\frac{\hbar}{i} \underset{S}{\leftrightarrow} x_k\left[-m\left(\frac{x_{k+1}-x_k}{\epsilon} - \frac{x_k-x_{k-1}}{\epsilon}\right) - \epsilon V'(x_k)\right]$$

or, neglecting terms of order ϵ,

$$m\left(\frac{x_{k+1}-x_k}{\epsilon}\right)x_k - m\left(\frac{x_k-x_{k-1}}{\epsilon}\right)x_k \underset{S}{\leftrightarrow} \frac{\hbar}{i}. \quad (49)$$

In order to transfer an equation such as (49) into conventional notation, we shall have to discover what matrix corresponds to a quantity such as $x_k x_{k+1}$. It is clear from a study of (39) that if F is set equal to, say, $f(x_k)g(x_{k+1})$, the corresponding operator in (40) is

$$e^{-(i/\hbar)(t''-t-\epsilon)\mathbf{H}}g(\mathbf{x})e^{-(i/\hbar)\epsilon\mathbf{H}}f(\mathbf{x})e^{-(i/\hbar)(t-t')\mathbf{H}},$$

the matrix element being taken between the states $\chi_{t''}$ and $\psi_{t'}$. The operators corresponding to functions of x_{k+1} will appear to the left of the operators corresponding to functions of x_k, i.e., *the order of terms in a matrix operator product corresponds to an order in time of the corresponding factors in a functional.* Thus, if the functional can and is written in such a way that in each term factors corresponding to later times appear to the left of factors corresponding to earlier terms, the corresponding operator can immediately be written down if the order of the operators is kept the same as in the functional.[18] Obviously, the order of factors in a functional is of no consequence. The ordering just facilitates translation into conventional operator notation. To write Eq. (49) in the way desired for easy translation would require the factors in the second term on the left to be reversed in order. We see, therefore, that it corresponds to

$$\mathbf{px} - \mathbf{xp} = \hbar/i$$

where we have written $\mathbf{p}$ for the operator $m\dot{x}$.

The relation between functionals and the corresponding operators is defined above in terms of the order of the factors in time. It should be remarked that this rule must be especially carefully adhered to when quantities involving velocities or higher derivatives are involved. The correct functional to represent the operator $(\dot{x})^2$ is actually $(x_{k+1}-x_k)/\epsilon \cdot (x_k-x_{k-1})/\epsilon$ rather than $[(x_{k+1}-x_k)/\epsilon]^2$. The latter quantity diverges as $1/\epsilon$ as $\epsilon \to 0$. This may be seen by replacing the second term in (49) by its value $x_{k+1} \cdot m(x_{k+1}-x_k)/\epsilon$ calculated an instant ϵ later in time. This does not change the equation to zero order in ϵ. We then obtain (dividing by ϵ)

$$\left(\frac{x_{k+1}-x_k}{\epsilon}\right)^2 \underset{s}{\leftrightarrow} \frac{\hbar}{im\epsilon}. \qquad (50)$$

This gives the result expressed earlier that the root mean square of the "velocity" $(x_{k+1}-x_k)/\epsilon$ between two successive positions of the path is of order $\epsilon^{-\frac{1}{2}}$.

It will not do then to write the functional for kinetic energy, say, simply as

$$\tfrac{1}{2}m[(x_{k+1}-x_k)/\epsilon]^2 \qquad (51)$$

for this quantity is infinite as $\epsilon \to 0$. In fact, it is not an observable functional.

One can obtain the kinetic energy as an observable functional by considering the first-order change in transition amplitude occasioned by a change in the mass of the particle. Let m be changed to $m(1+\delta)$ for a short time, say ϵ, around t_k. The change in the action is $\tfrac{1}{2}\delta\epsilon m[(x_{k+1}-x_k)/\epsilon]^2$

[18] Dirac has also studied operators containing quantities referring to different times. See reference 2.

the derivative of which gives an expression like (51). But the change in m changes the normalization constant $1/A$ corresponding to dx_k as well as the action. The constant is changed from $(2\pi\hbar\epsilon i/m)^{-\frac{1}{2}}$ to $(2\pi\hbar\epsilon i/m(1+\delta))^{-\frac{1}{2}}$ or by $\tfrac{1}{2}\delta(2\pi\hbar\epsilon i/m)^{-\frac{1}{2}}$ to first order in δ. The total effect of the change in mass in Eq. (38) to the first order in δ is

$$\langle \chi_{t''} | \tfrac{1}{2}\delta\epsilon im[(x_{k+1}-x_k)/\epsilon]^2/\hbar + \tfrac{1}{2}\delta | \psi_{t'} \rangle.$$

We expect the change of order δ lasting for a time ϵ to be of order $\delta\epsilon$. Hence, dividing by $\delta\epsilon i/\hbar$, we can define the kinetic energy functional as

$$\text{K.E.} = \tfrac{1}{2}m[(x_{k+1}-x_k)/\epsilon]^2 + \hbar/2\epsilon i. \qquad (52)$$

This is finite as $\epsilon \to 0$ in view of (50). By making use of an equation which results from substituting $m(x_{k+1}-x_k)/\epsilon$ for F in (48) we can also show that the expression (52) is equal (to order ϵ) to

$$\text{K.E.} = \tfrac{1}{2}m\left(\frac{x_{k+1}-x_k}{\epsilon}\right)\left(\frac{x_k-x_{k-1}}{\epsilon}\right). \qquad (53)$$

That is, the easiest way to produce observable functionals involving powers of the velocities is to replace these powers by a product of velocities, each factor of which is taken at a slightly different time.

10. THE HAMILTONIAN

Momentum

The Hamiltonian operator is of central importance in the usual formulation of quantum mechanics. We shall study in this section the functional corresponding to this operator. We could immediately define the Hamiltonian functional by adding the kinetic energy functional (52) or (53) to the potential energy. This method is artificial and does not exhibit the important relationship of the Hamiltonian to time. We shall define the Hamiltonian functional by the changes made in a state when it is displaced in time.

To do this we shall have to digress a moment to point out that the subdivision of time into *equal* intervals is not necessary. Clearly, any subdivision into instants t_i will be satisfactory; the limits are to be taken as the largest spacing, $t_{i+1}-t_i$, approaches zero. The total action S must now be represented as a sum

$$S = \sum_i S(x_{i+1}, t_{i+1}; x_i, t_i), \quad (54)$$

where

$$S(x_{i+1}, t_{i+1}; x_i, t_i) = \int_{t_i}^{t_{i+1}} L(\dot{x}(t), x(t))dt, \quad (55)$$

the integral being taken along the classical path between x_i at t_i and x_{i+1} at t_{i+1}. For the simple one-dimensional example this becomes, with sufficient accuracy,

$$S(x_{i+1}, t_{i+1}; x_i, t_i)$$
$$= \left\{ \frac{m}{2}\left(\frac{x_{i+1}-x_i}{t_{i+1}-t_i}\right)^2 - V(x_{i+1}) \right\}(t_{i+1}-t_i); \quad (56)$$

the corresponding normalization constant for integration on dx_i is $A = (2\pi h i(t_{i+1}-t_i)/m)^{-\frac{1}{2}}$.

The relation of H to the change in a state with displacement in time can now be studied. Consider a state $\psi(t)$ defined by a space-time region R'. Now imagine that we consider another state at time t, $\psi_\delta(t)$, defined by another region R_δ'. Suppose the region R_δ' is exactly the same as R' except that it is earlier by a time δ, i.e., displaced bodily toward the past by a time δ. All the apparatus to prepare the system for R_δ' is identical to that for R' but is operated a time δ sooner. If L depends explicitly on time, it, too, is to be displaced, i.e., the state ψ_δ is obtained from the L used for state ψ except that the time t in L_δ is replaced by $t+\delta$. We ask how does the state ψ_δ differ from ψ? In any measurement the chance of finding the system in a fixed region R'' is different for R' and R_δ'. Consider the change in the transition element $\langle \chi | 1 | \psi_\delta \rangle_{S_\delta}$ produced by the shift δ. We can consider this shift as effected by decreasing all values of t_i by δ for $i \leqslant k$ and leaving all t_i fixed for $i > k$, where the time t lies in the interval between t_{k+1} and t_k.[19] This change will have no effect on $S(x_{i+1}, t_{i+1}; x_i, t_i)$ as defined by (55) as long as both t_{i+1} and t_i are changed by the same amount. On the other hand, $S(x_{k+1}, t_{k+1}; x_k, t_k)$

is changed to $S(x_{k+1}, t_{k+1}; x_k, t_k-\delta)$. The constant $1/A$ for the integration on dx_k is also altered to $(2\pi h i(t_{k+1}-t_k+\delta)/m)^{-\frac{1}{2}}$. The effect of these changes on the transition element is given to the first order in δ by

$$\langle \chi | 1 | \psi \rangle_S - \langle \chi | 1 | \psi_\delta \rangle_{S_\delta} = \frac{i\delta}{\hbar}\langle \chi | H_k | \psi \rangle_S, \quad (57)$$

here the Hamiltonian functional H_k is defined by

$$H_k = \frac{\partial S(x_{k+1}, t_{k+1}; x_k, t_k)}{\partial t_k} + \frac{\hbar}{2i(t_{k+1}-t_k)}. \quad (58)$$

The last term is due to the change in $1/A$ and serves to keep H_k finite as $\epsilon \to 0$. For example, for the expression (56) this becomes

$$H_k = \frac{m}{2}\left(\frac{x_{k+1}-x_k}{t_{k+1}-t_k}\right)^2 + \frac{\hbar}{2i(t_{k+1}-t_k)} + V(x_{k+1}),$$

which is just the sum of the kinetic energy functional (52) and that of the potential energy $V(x_{k+1})$.

The wave function $\psi_\delta(x,t)$ represents, of course, the same state as $\psi(x,t)$ will be after time δ, i.e., $\psi(x, t+\delta)$. Hence, (57) is intimately related to the operator equation (31).

One could also consider changes occasioned by a time shift in the final state χ. Of course, nothing new results in this way for it is only the relative shift of χ and ψ which counts. One obtains an alternative expression

$$H_k = -\frac{\partial S(x_{k+1}, t_{k+1}; x_k, t_k)}{\partial t_{k+1}} + \frac{\hbar}{2i(t_{k+1}-t_k)}. \quad (59)$$

This differs from (58) only by terms of order ϵ.

The time rate of change of a functional can be computed by considering the effect of shifting both initial and final state together. This has the same effect as calculating the transition element of the functional referring to a later time. What results is the analog of the operator equation

$$\frac{\hbar}{i}\dot{\mathbf{f}} = \mathbf{Hf} - \mathbf{fH}.$$

The momentum functional p_k can be defined in an analagous way by considering the changes

---

[19] From the point of view of mathematical rigor, if δ is finite, as $\epsilon \to 0$ one gets into difficulty in that, for example, the interval $t_{k+1}-t_k$ is kept finite. This can be straightened out by assuming δ to vary with time and to be turned on smoothly before $t=t_k$ and turned off smoothly after $t=t_k$. Then keeping the time variation of δ fixed, let $\epsilon \to 0$. Then seek the first-order change as $\delta \to 0$. The result is essentially the same as that of the crude procedure used above.

made by displacements of position:

$$\langle \chi|1|\psi\rangle_S - \langle \chi|1|\psi_\Delta\rangle_{S_\Delta} = \frac{i\Delta}{\hbar}\langle \chi|p_k|\psi\rangle_S.$$

The state ψ_Δ is prepared from a region R_Δ' which is identical to region R' except that it is moved a distance Δ in space. (The Lagrangian, if it depends explicitly on x, must be altered to $L_\Delta = L(\dot{x}, x-\Delta)$ for times previous to t.) One finds[20]

$$p_k = \frac{\partial S(x_{k+1}, x_k)}{\partial x_{k+1}} = -\frac{\partial S(x_{k+1}, x_k)}{\partial x_k}. \quad (60)$$

Since $\psi_\Delta(x,t)$ is equal to $\psi(x-\Delta, t)$, the close connection between p_k and the x-derivative of the wave function is established.

Angular momentum operators are related in an analogous way to rotations.

The derivative with respect to t_{i+1} of $S(x_{i+1}, t_{i+1}; x_i, t_i)$ appears in the definition of H_i. The derivative with respect to x_{i+1} defines p_i. But the derivative with respect to t_{i+1} of $S(x_{i+1}, t_{i+1}; x_i, t_i)$ is related to the derivative with respect to x_{i+1}, for the function $S(x_{i+1}, t_{i+1}; x_i, t_i)$ defined by (55) satisfies the Hamilton-Jacobi equation. Thus, the Hamilton-Jacobi equation is an equation expressing H_i in terms of the p_i. In other words, it expresses the fact that time displacements of states are related to space displacements of the same states. This idea leads directly to a derivation of the Schroedinger equation which is far more elegant than the one exhibited in deriving Eq. (30).

11. INADEQUACIES OF THE FORMULATION

The formulation given here suffers from a serious drawback. The mathematical concepts needed are new. At present, it requires an unnatural and cumbersome subdivision of the time interval to make the meaning of the equations clear. Considerable improvement can be made through the use of the notation and concepts of the mathematics of functionals. However, it was thought best to avoid this in a first presentation. One

[20] We did not immediately substitute p_i from (60) into (47) because (47) would then no longer have been valid to both zero order and the first order in ϵ. We could derive the commutation relations, but not the equations of motion. The two expressions in (60) represent the momenta at each end of the interval t_i to t_{i+1}. They differ by $\epsilon V'(x_{k+1})$ because of the force acting during the time ϵ.

needs, in addition, an appropriate measure for the space of the argument functions $x(t)$ of the functionals.[10]

It is also incomplete from the physical standpoint. One of the most important characteristics of quantum mechanics is its invariance under unitary transformations. These correspond to the canonical transformations of classical mechanics. Of course, the present formulation, being equivalent to ordinary formulations, can be mathematically demonstrated to be invariant under these transformations. However, it has not been formulated in such a way that it is *physically* obvious that it is invariant. This incompleteness shows itself in a definite way. No direct procedure has been outlined to describe measurements of quantities other than position. Measurements of momentum, for example, of one particle, can be defined in terms of measurements of positions of other particles. The result of the analysis of such a situation does show the connection of momentum measurements to the Fourier transform of the wave function. But this is a rather roundabout method to obtain such an important physical result. It is to be expected that the postulates can be generalized by the replacement of the idea of "paths in a region of space-time R" to "paths of class R," or "paths having property R." But which properties correspond to which physical measurements has not been formulated in a general way.

12. A POSSIBLE GENERALIZATION

The formulation suggests an obvious generalization. There are interesting classical problems which satisfy a principle of least action but for which the action cannot be written as an integral of a function of positions and velocities. The action may involve accelerations, for example. Or, again, if interactions are not instantaneous, it may involve the product of coordinates at two different times, such as $\int x(t)x(t+T)dt$. The action, then, cannot be broken up into a sum of small contributions as in (10). As a consequence, no wave function is available to describe a state. Nevertheless, a transition probability can be defined for getting from a region R' into another R''. Most of the theory of the transition elements $\langle \chi_{t''}|F|\psi_{t'}\rangle_S$ can be carried over. One simply invents a symbol, such as $\langle R''|F|R'\rangle_S$ by an equation such as (39) but with the expressions (19) and (20) for ψ and χ substituted, and the more general action substituted for S. Hamiltonian and momentum functionals can be defined as in section (10). Further details may be found in a thesis by the author.[21]

13. APPLICATION TO ELIMINATE FIELD OSCILLATORS

One characteristic of the present formulation is that it can give one a sort of bird's-eye view of the space-time relationships in a given situation. Before the integrations on the x_i are performed in an expression such as (39) one has a sort of format into which various F functionals may be inserted. One can study how what goes on in the quantum-mechanical system at different times is interrelated. To make these vague remarks somewhat more definite, we discuss an example.

In classical electrodynamics the fields describing, for instance, the interaction of two particles can be represented as a set of oscillators. The equations of motion of these oscillators may be solved and the oscillators essentially eliminated (Lienard and Wiechert potentials). The interactions which result involve relationships of the motion of one particle at one time, and of the other particle at another time. In quantum electrodynamics the field is again represented as a set of oscillators. But the motion of the oscillators cannot be worked out and the oscillators eliminated. It is true that the oscillators representing longitudinal waves may be eliminated. The result is instantaneous electrostatic interaction. The electrostatic elimination is very instructive as it shows up the difficulty of self-interaction very distinctly. In fact, it shows it up so clearly that there is no ambiguity in deciding what term is incorrect and should be omitted. This entire process is not relativistically invariant, nor is the omitted term. It would seem to be very desirable if the oscillators, representing transverse waves,

[21] The theory of electromagnetism described by J. A. Wheeler and R. P. Feynman, Rev. Mod. Phys. **17**, 157 (1945) can be expressed in a principle of least action involving the coordinates of particles alone. It was an attempt to quantize this theory, without reference to the fields, which led the author to study the formulation of quantum mechanics given here. The extension of the ideas to cover the case of more general action functions was developed in his Ph.D. thesis, "The principle of least action in quantum mechanics" submitted to Princeton University, 1942.

could also be eliminated. This presents an almost insurmountable problem in the conventional quantum mechanics. We expect that the motion of a particle a at one time depends upon the motion of b at a previous time, and *vice versa*. A wave function $\psi(x_a, x_b; t)$, however, can only describe the behavior of both particles at one time. There is no way to keep track of what b did in the past in order to determine the behavior of a. The only way is to specify the state of the set of oscillators at t, which serve to "remember" what b (and a) had been doing.

The present formulation permits the solution of the motion of all the oscillators and their complete elimination from the equations describing the particles. This is easily done. One must simply solve for the motion of the oscillators before one integrates over the various variables x_i for the particles. It is the integration over x_i which tries to condense the past history into a single state function. This we wish to avoid. Of course, the result depends upon the initial and final states of the oscillator. If they are specified, the result is an equation for $\langle \chi_{t''} | 1 | \psi_{t'} \rangle$ like (38), but containing as a factor, besides $\exp(iS/\hbar)$ another functional G depending only on the coordinates describing the paths of the particles.

We illustrate briefly how this is done in a very simple case. Suppose a particle, coordinate $x(t)$, Lagrangian $L(\dot{x}, x)$ interacts with an oscillator, coordinate $q(t)$, Lagrangian $\frac{1}{2}(\dot{q}^2 - \omega^2 q^2)$, through a term $\gamma(x, t)q(t)$ in the Lagrangian for the system. Here $\gamma(x, t)$ is any function of the coordinate $x(t)$ of the particle and the time.[22] Suppose we desire the probability of a transition from a state at time t', in which the particle's wave function is $\psi_{t'}$ and the oscillator is in energy level n, to a state at t'' with the particle in $\chi_{t''}$ and oscillator in level m. This is the square of

$$\langle \chi_{t''} \varphi_m | 1 | \psi_{t'} \varphi_n \rangle_{S_p + S_0 + S_I}$$

$$= \int \cdots \int \varphi_m{}^*(q_j) \chi_{t''}{}^*(x_j)$$
$$\times \exp\frac{i}{\hbar}(S_p + S_0 + S_I) \psi_{t'}(x_0) \varphi_n(q_0)$$
$$\frac{dx_0}{A}\frac{dq_0}{a} \cdots \frac{dx_{j-1}}{A}\frac{dq_{j-1}}{a} dx_j dq_j. \quad (61)$$

[22] The generalization to the case that γ depends on the velocity, $\dot{x}$, of the particle presents no problem.

Here $\varphi_n(q)$ is the wave function for the oscillator in state n, S_p is the action

$$\sum_{i=0}^{j-1} S_p(x_{i+1}, x_i)$$

calculated for the particle as though the oscillator were absent,

$$S_0 = \sum_{i=0}^{j-1}\left[\frac{\epsilon}{2}\left(\frac{q_{i+1}-q_i}{\epsilon}\right)^2 - \frac{\epsilon\omega^2}{2}q_{i+1}^2\right]$$

that of the oscillator alone, and

$$S_I = \sum_{i=0}^{j-1} \gamma_i q_i$$

(where $\gamma_i = \gamma(x_i, t_i)$) is the action of interaction between the particle and the oscillator. The normalizing constant, a, for the oscillator is $(2\pi\epsilon i/\hbar)^{-\frac{1}{2}}$. Now the exponential depends quadratically upon all the q_i. Hence, the integrations over all the variables q_i for $0 < i < j$ can easily be performed. One is integrating a sequence of Gaussian integrals.

The result of these integrations is, writing $T = t'' - t'$, $(2\pi i\hbar \sin\omega T/\omega)^{-\frac{1}{2}} \exp i(S_p + Q(q_j, q_0))/\hbar$, where $Q(q_j, q_0)$ turns out to be just the classical action for the forced harmonic oscillator (see reference 15). Explicitly it is

$$Q(q_j, q_0) = \frac{\omega}{2 \sin\omega T}\Big[(\cos\omega T)(q_j^2 + q_0^2) - 2q_j q_0$$
$$+ \frac{2q_0}{\omega}\int_{t'}^{t''} \gamma(t) \sin\omega(t-t')dt$$
$$+ \frac{2q_j}{\omega}\int_{t'}^{t''} \gamma(t) \sin\omega(t''-t)dt$$
$$- \frac{2}{\omega^2}\int_{t'}^{t''}\int_{t'}^{t} \gamma(t)\gamma(s) \sin\omega(t''-t)$$
$$\times \sin\omega(s-t')dsdt\Big].$$

It has been written as though $\gamma(t)$ were a continuous function of time. The integrals really should be split into Riemann sums and the quantity $\gamma(x_i, t_i)$ substituted for $\gamma(t_i)$. Thus, Q depends on the coordinates of the particle at all times through the $\gamma(x_i, t_i)$ and on that of the oscillator at times t' and t'' only. Thus, the quantity (61) becomes

$$\langle \chi_{t''} \varphi_m | 1 | \psi_{t'} \varphi_n \rangle_{S_p + S_0 + S_I} = \int \cdots \int \chi_{t''}*(x_j)G_{mn}$$
$$\times \exp\left(\frac{iS_p}{\hbar}\right)\psi_{t'}(x_0)\frac{dx_0}{A}\cdots\frac{dx_{j-1}}{A}dx_j$$
$$= \langle \chi_{t''} | G_{mn} | \psi_{t'} \rangle_{S_p}$$

which now contains the coordinates of the particle only, the quantity G_{mn} being given by

$$G_{mn} = (2\pi i\hbar \sin\omega T/\omega)^{-\frac{1}{2}} \int\int \varphi_m*(q_j)$$
$$\times \exp(iQ(q_j, q_0)/\hbar)\varphi_n(q_0)dq_j dq_0.$$

Proceeding in an analogous manner one finds that all of the oscillators of the electromagnetic field can be eliminated from a description of the motion of the charges.

14. STATISTICAL MECHANICS
Spin and Relativity

Problems in the theory of measurement and statistical quantum mechanics are often simplified when set up from the point of view described here. For example, the influence of a perturbing measuring instrument can be integrated out in principle as we did in detail for the oscillator. The statistical density matrix has a fairly obvious and useful generalization. It results from considering the square of (38). It is an expression similar to (38) but containing integrations over two sets of variables dx_i and dx_i'. The exponential is replaced by $\exp i(S-S')/\hbar$, where S' is the same function of the x_i' as S is of x_i. It is required, for example, to describe the result of the elimination of the field oscillators where, say, the final state of the oscillators is unspecified and one desires only the sum over all final states m.

Spin may be included in a formal way. The Pauli spin equation can be obtained in this way:

One replaces the vector potential interaction term in $S(x_{i+1}, x_i)$,

$$\frac{e}{2c}(\mathbf{x}_{i+1}-\mathbf{x}_i)\cdot\mathbf{A}(\mathbf{x}_i)+\frac{e}{2c}(\mathbf{x}_{i+1}-\mathbf{x}_i)\cdot\mathbf{A}(\mathbf{x}_{i+1})$$

arising from expression (13) by the expression

$$\frac{e}{2c}(\boldsymbol{\sigma}\cdot(\mathbf{x}_{i+1}-\mathbf{x}_i))(\boldsymbol{\sigma}\cdot\mathbf{A}(\mathbf{x}_i))$$
$$+\frac{e}{2c}(\boldsymbol{\sigma}\cdot\mathbf{A}(\mathbf{x}_{i+1}))(\boldsymbol{\sigma}\cdot(\mathbf{x}_{i+1}-\mathbf{x}_i)).$$

Here $\mathbf{A}$ is the vector potential, $\mathbf{x}_{i+1}$ and $\mathbf{x}_i$ the vector positions of a particle at times t_{i+1} and t_i and $\boldsymbol{\sigma}$ is Pauli's spin vector matrix. The quantity Φ must now be expressed as $\prod_i \exp iS(x_{i+1}, x_i)/\hbar$ for this differs from the exponential of the sum of $S(x_{i+1}, x_i)$. Thus, Φ is now a spin matrix.

The Klein Gordon relativistic equation can also be obtained formally by adding a fourth coordinate to specify a path. One considers a "path" as being specified by four functions $x^{(\mu)}(\tau)$ of a parameter τ. The parameter τ now goes in steps ϵ as the variable t went previously. The quantities $x^{(1)}(t)$, $x^{(2)}(t)$, $x^{(3)}(t)$ are the space coordinates of a particle and $x^{(4)}(t)$ is a corresponding time. The Lagrangian used is

$$\sum_{\mu=1}^{4}{}' [(dx^\mu/d\tau)^2+(e/c)(dx^\mu/d\tau)A_\mu],$$

where A_μ is the 4-vector potential and the terms in the sum for $\mu=1, 2, 3$ are taken with reversed sign. If one seeks a wave function which depends upon τ periodically, one can show this must satisfy the Klein Gordon equation. The Dirac equation results from a modification of the Lagrangian used for the Klein Gordon equation, which is analogous to the modification of the non-relativistic Lagrangian required for the Pauli equation. What results directly is the square of the usual Dirac operator.

These results for spin and relativity are purely formal and add nothing to the understanding of these equations. There are other ways of obtaining the Dirac equation which offer some promise of giving a clearer physical interpretation to that important and beautiful equation.

The author sincerely appreciates the helpful advice of Professor and Mrs. H. C. Corben and of Professor H. A. Bethe. He wishes to thank Professor J. A. Wheeler for very many discussions during the early stages of the work.

논문 웹페이지

Space-Time Approach to Quantum Electrodynamics

R. P. FEYNMAN
Department of Physics, Cornell University, Ithaca, New York
(Received May 9, 1949)

In this paper two things are done. (1) It is shown that a considerable simplification can be attained in writing down matrix elements for complex processes in electrodynamics. Further, a physical point of view is available which permits them to be written down directly for any specific problem. Being simply a restatement of conventional electrodynamics, however, the matrix elements diverge for complex processes. (2) Electrodynamics is modified by altering the interaction of electrons at short distances. All matrix elements are now finite, with the exception of those relating to problems of vacuum polarization. The latter are evaluated in a manner suggested by Pauli and Bethe, which gives finite results for these matrices also. The only effects sensitive to the modification are changes in mass and charge of the electrons. Such changes could not be directly observed. Phenomena directly observable, are insensitive to the details of the modification used (except at extreme energies). For such phenomena, a limit can be taken as the range of the modification goes to zero. The results then agree with those of Schwinger. A complete, unambiguous,

and presumably consistent, method is therefore available for the calculation of all processes involving electrons and photons.

The simplification in writing the expressions results from an emphasis on the over-all space-time view resulting from a study of the solution of the equations of electrodynamics. The relation of this to the more conventional Hamiltonian point of view is discussed. It would be very difficult to make the modification which is proposed if one insisted on having the equations in Hamiltonian form.

The methods apply as well to charges obeying the Klein-Gordon equation, and to the various meson theories of nuclear forces. Illustrative examples are given. Although a modification like that used in electrodynamics can make all matrices finite for all of the meson theories, for some of the theories it is no longer true that all directly observable phenomena are insensitive to the details of the modification used.

The actual evaluation of integrals appearing in the matrix elements may be facilitated, in the simpler cases, by methods described in the appendix.

THIS paper should be considered as a direct continuation of a preceding one[1] (I) in which the motion of electrons, neglecting interaction, was analyzed, by dealing directly with the *solution* of the Hamiltonian differential equations. Here the same technique is applied to include interactions and in that way to express in simple terms the solution of problems in quantum electrodynamics.

For most practical calculations in quantum electrodynamics the solution is ordinarily expressed in terms of a matrix element. The matrix is worked out as an expansion in powers of $e^2/\hbar c$, the successive terms corresponding to the inclusion of an increasing number of virtual quanta. It appears that a considerable simplification can be achieved in writing down these matrix elements for complex processes. Furthermore, each term in the expansion can be written down and understood directly from a physical point of view, similar to the space-time view in I. It is the purpose of this paper to describe how this may be done. We shall also discuss methods of handling the divergent integrals which appear in these matrix elements.

The simplification in the formulae results mainly from the fact that previous methods unnecessarily separated into individual terms processes that were closely related physically. For example, in the exchange of a quantum between two electrons there were two terms depending on which electron emitted and which absorbed the quantum. Yet, in the virtual states considered, timing relations are not significant. Olny the order of operators in the matrix must be maintained. We have seen (I), that in addition, processes in which virtual pairs are produced can be combined with others in which only

positive energy electrons are involved. Further, the effects of longitudinal and transverse waves can be combined together. The separations previously made were on an unrelativistic basis (reflected in the circumstance that apparently momentum but not energy is conserved in intermediate states). When the terms are combined and simplified, the relativistic invariance of the result is self-evident.

We begin by discussing the solution in space and time of the Schrödinger equation for particles interacting instantaneously. The results are immediately generalizable to delayed interactions of relativistic electrons and we represent in that way the laws of quantum electrodynamics. We can then see how the matrix element for any process can be written down directly. In particular, the self-energy expression is written down.

So far, nothing has been done other than a restatement of conventional electrodynamics in other terms. Therefore, the self-energy diverges. A modification[2] in interaction between charges is next made, and it is shown that the self-energy is made convergent and corresponds to a correction to the electron mass. After the mass correction is made, other real processes are finite and insensitive to the "width" of the cut-off in the interaction.[3]

Unfortunately, the modification proposed is not completely satisfactory theoretically (it leads to some difficulties of conservation of energy). It does, however, seem consistent and satisfactory to define the matrix

---

[1] R. P. Feynman, Phys. Rev. **76**, 749 (1949), hereafter called I.

[2] For a discussion of this modification in classical physics see R. P. Feynman, Phys. Rev. **74** 939 (1948), hereafter referred to as A.

[3] A brief summary of the methods and results will be found in R. P. Feynman, Phys. Rev. **74**, 1430 (1948), hereafter referred to as B.

element for all real processes as the limit of that computed here as the cut-off width goes to zero. A similar technique suggested by Pauli and by Bethe can be applied to problems of vacuum polarization (resulting in a renormalization of charge) but again a strict physical basis for the rules of convergence is not known.

After mass and charge renormalization, the limit of zero cut-off width can be taken for all real processes. The results are then equivalent to those of Schwinger[4] who does not make explicit use of the convergence factors. The method of Schwinger is to identify the terms corresponding to corrections in mass and charge and, previous to their evaluation, to remove them from the expressions for real processes. This has the advantage of showing that the results can be strictly independent of particular cut-off methods. On the other hand, many of the properties of the integrals are analyzed using formal properties of invariant propagation functions. But one of the properties is that the integrals are infinite and it is not clear to what extent this invalidates the demonstrations. A practical advantage of the present method is that ambiguities can be more easily resolved; simply by direct calculation of the otherwise divergent integrals. Nevertheless, it is not at all clear that the convergence factors do not upset the physical consistency of the theory. Although in the limit the two methods agree, neither method appears to be thoroughly satisfactory theoretically. Nevertheless, it does appear that we now have available a complete and definite method for the calculation of physical processes to any order in quantum electrodynamics.

Since we can write down the solution to any physical problem, we have a complete theory which could stand by itself. It will be theoretically incomplete, however, in two respects. First, although each term of increasing order in $e^2/\hbar c$ can be written down it would be desirable to see some way of expressing things in finite form to all orders in $e^2/\hbar c$ at once. Second, although it will be physically evident that the results obtained are equivalent to those obtained by conventional electrodynamics the mathematical proof of this is not included. Both of these limitations will be removed in a subsequent paper (see also Dyson[4]).

Briefly the genesis of this theory was this. The conventional electrodynamics was expressed in the Lagrangian form of quantum mechanics described in the Reviews of Modern Physics.[5] The motion of the field oscillators could be integrated out (as described in Section 13 of that paper), the result being an expression of the delayed interaction of the particles. Next the modification of the delta-function interaction could be made directly from the analogy to the classical case.[2] This was still not complete because the Lagrangian method had been worked out in detail only for particles obeying the non-relativistic Schrödinger equation. It was then modified in accordance with the requirements of the Dirac equation and the phenomenon of pair creation. This was made easier by the reinterpretation of the theory of holes (I). Finally for practical calculations the expressions were developed in a power series in $e^2/\hbar c$. It was apparent that each term in the series had a simple physical interpretation. Since the result was easier to understand than the derivation, it was thought best to publish the results first in this paper. Considerable time has been spent to make these first two papers as complete and as physically plausible as possible without relying on the Lagrangian method, because it is not generally familiar. It is realized that such a description cannot carry the conviction of truth which would accompany the derivation. On the other hand, in the interest of keeping simple things simple the derivation will appear in a separate paper.

The possible application of these methods to the various meson theories is discussed briefly. The formulas corresponding to a charge particle of zero spin moving in accordance with the Klein Gordon equation are also given. In an Appendix a method is given for calculating the integrals appearing in the matrix elements for the simpler processes.

The point of view which is taken here of the interaction of charges differs from the more usual point of view of field theory. Furthermore, the familiar Hamiltonian form of quantum mechanics must be compared to the over-all space-time view used here. The first section is, therefore, devoted to a discussion of the relations of these viewpoints.

1. COMPARISON WITH THE HAMILTONIAN METHOD

Electrodynamics can be looked upon in two equivalent and complementary ways. One is as the description of the behavior of a field (Maxwell's equations). The other is as a description of a direct interaction at a distance (albeit delayed in time) between charges (the solutions of Lienard and Wiechert). From the latter point of view light is considered as an interaction of the charges in the source with those in the absorber. This is an impractical point of view because many kinds of sources produce the same kind of effects. The field point of view separates these aspects into two simpler problems, production of light, and absorption of light. On the other hand, the field point of view is less practical when dealing with close collisions of particles (or their action on themselves). For here the source and absorber are not readily distinguishable, there is an intimate exchange of quanta. The fields are so closely determined by the motions of the particles that it is just as well not to separate the question into two problems but to consider the process as a direct interaction. Roughly, the field point of view is most practical for problems involv-

---

[4] J. Schwinger, Phys. Rev. **74**, 1439 (1948), Phys. Rev. **75**, 651 (1949). A proof of this equivalence is given by F. J. Dyson, Phys. Rev. **75**, 486 (1949).

[5] R. P. Feynman, Rev. Mod. Phys. **20**, 367 (1948). The application to electrodynamics is described in detail by H. J. Groenewold, Koninklijke Nederlandsche Akademia van Weteschappen. Proceedings Vol. LII, 3 (226) 1949.

ing real quanta, while the interaction view is best for the discussion of the virtual quanta involved. We shall emphasize the interaction viewpoint in this paper, first because it is less familiar and therefore requires more discussion, and second because the important aspect in the problems with which we shall deal is the effect of virtual quanta.

The Hamiltonian method is not well adapted to represent the direct action at a distance between charges because that action is delayed. The Hamiltonian method represents the future as developing out of the present. If the values of a complete set of quantities are known now, their values can be computed at the next instant in time. If particles interact through a delayed interaction, however, one cannot predict the future by simply knowing the present motion of the particles. One would also have to know what the motions of the particles were in the past in view of the interaction this may have on the future motions. This is done in the Hamiltonian electrodynamics, of course, by requiring that one specify besides the present motion of the particles, the values of a host of new variables (the coordinates of the field oscillators) to keep track of that aspect of the past motions of the particles which determines their future behavior. The use of the Hamiltonian forces one to choose the field viewpoint rather than the interaction viewpoint.

In many problems, for example, the close collisions of particles, we are not interested in the precise temporal sequence of events. It is not of interest to be able to say how the situation would look at each instant of time during a collision and how it progresses from instant to instant. Such ideas are only useful for events taking a long time and for which we can readily obtain information during the intervening period. For collisions it is much easier to treat the process as a whole.[6] The Møller interaction matrix for the the collision of two electrons is not essentially more complicated than the nonrelativistic Rutherford formula, yet the mathematical machinery used to obtain the former from quantum electrodynamics is vastly more complicated than Schrödinger's equation with the e^2/r_{12} interaction needed to obtain the latter. The difference is only that in the latter the action is instantaneous so that the Hamiltonian method requires no extra variables, while in the former relativistic case it is delayed and the Hamiltonian method is very cumbersome.

We shall be discussing the solutions of equations rather than the time differential equations from which they come. We shall discover that the solutions, because of the over-all space-time view that they permit, are as easy to understand when interactions are delayed as when they are instantaneous.

As a further point, relativistic invariance will be self-evident. The Hamiltonian form of the equations develops the future from the instantaneous present. But

---

[6] This is the viewpoint of the theory of the S matrix of Heisenberg.

for different observers in relative motion the instantaneous present is different, and corresponds to a different 3-dimensional cut of space-time. Thus the temporal analyses of different observers is different and their Hamiltonian equations are developing the process in different ways. These differences are irrelevant, however, for the solution is the same in any space time frame. By forsaking the Hamiltonian method, the wedding of relativity and quantum mechanics can be accomplished most naturally.

We illustrate these points in the next section by studying the solution of Schrödinger's equation for nonrelativistic particles interacting by an instantaneous Coulomb potential (Eq. 2). When the solution is modified to include the effects of delay in the interaction and the relativistic properties of the electrons we obtain an expression of the laws of quantum electrodynamics (Eq. 4).

2. THE INTERACTION BETWEEN CHARGES

We study by the same methods as in I, the interaction of two particles using the same notation as I. We start by considering the non-relativistic case described by the Schrödinger equation (I, Eq. 1). The wave function at a given time is a function $\psi(\mathbf{x}_a, \mathbf{x}_b, t)$ of the coordinates $\mathbf{x}_a$ and $\mathbf{x}_b$ of each particle. Thus call $K(\mathbf{x}_a, \mathbf{x}_b, t; \mathbf{x}_a', \mathbf{x}_b', t')$ the amplitude that particle a at $\mathbf{x}_a'$ at time t' will get to $\mathbf{x}_a$ at t while particle b at $\mathbf{x}_b'$ at t' gets to $\mathbf{x}_b$ at t. If the particles are free and do not interact this is

$$K(\mathbf{x}_a, \mathbf{x}_b, t; \mathbf{x}_a', \mathbf{x}_b', t') = K_{0a}(\mathbf{x}_a, t; \mathbf{x}_a', t')K_{0b}(\mathbf{x}_b, t; \mathbf{x}_b', t')$$

where K_{0a} is the K_0 function for particle a considered as free. In *this* case we can obviously define a quantity like K, but for which the time t need not be the same for particles a and b (likewise for t'); e.g.,

$$K_0(3, 4; 1, 2) = K_{0a}(3, 1)K_{0b}(4, 2) \quad (1)$$

can be thought of as the amplitude that particle a goes from $\mathbf{x}_1$ at t_1 to $\mathbf{x}_3$ at t_3 and that particle b goes from $\mathbf{x}_2$ at t_2 to $\mathbf{x}_4$ at t_4.

When the particles do interact, one can only define the quantity $K(3, 4; 1, 2)$ precisely if the interaction vanishes between t_1 and t_2 and also between t_3 and t_4. In a real physical system such is not the case. There is such an enormous advantage, however, to the concept that we shall continue to use it, imagining that we can neglect the effect of interactions between t_1 and t_2 and between t_3 and t_4. For practical problems this means choosing such long time intervals $t_3 - t_1$ and $t_4 - t_2$ that the extra interactions near the end points have small relative effects. As an example, in a scattering problem it may well be that the particles are so well separated initially and finally that the interaction at these times is negligible. Again energy values can be defined by the average rate of change of phase over such long time intervals that errors initially and finally can be neglected. Inasmuch as any physical problem can be defined in terms of scattering processes we do not lose much in

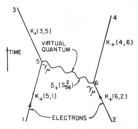

Fig. 1. The fundamental interaction Eq. (4). Exchange of one quantum between two electrons.

a general theoretical sense by this approximation. If it is not made it is not easy to study interacting particles relativistically, for there is nothing significant in choosing $t_1 = t_3$ if $x_1 \neq x_3$, as absolute simultaneity of events at a distance cannot be defined invariantly. It is essentially to avoid this approximation that the complicated structure of the older quantum electrodynamics has been built up. We wish to describe electrodynamics as a delayed interaction between particles. If we can make the approximation of assuming a meaning to $K(3,4;1,2)$ the results of this interaction can be expressed very simply.

To see how this may be done, imagine first that the interaction is simply that given by a Coulomb potential e^2/r where r is the distance between the particles. If this be turned on only for a very short time Δt_0 at time t_0, the first order correction to $K(3,4;1,2)$ can be worked out exactly as was Eq. (9) of I by an obvious generalization to two particles:

$$K^{(1)}(3,4;1,2) = -ie^2 \iint K_{0a}(3,5)K_{0b}(4,6)r_{56}^{-1}$$
$$\times K_{0a}(5,1)K_{0b}(6,2)d^3\mathbf{x}_5 d^3\mathbf{x}_6 \Delta t_0,$$

where $t_5 = t_6 = t_0$. If now the potential were on at all times (so that strictly K is not defined unless $t_4 = t_3$ and $t_1 = t_2$), the first-order effect is obtained by integrating on t_0, which we can write as an integral over both t_5 and t_6 if we include a delta-function $\delta(t_5 - t_6)$ to insure contribution only when $t_5 = t_6$. Hence, the first-order effect of interaction is (calling $t_5 - t_6 = t_{56}$):

$$K^{(1)}(3,4;1,2) = -ie^2 \iint K_{0a}(3,5)K_{0b}(4,6)r_{56}^{-1}$$
$$\times \delta(t_{56})K_{0a}(5,1)K_{0b}(6,2)d\tau_5 d\tau_6, \quad (2)$$

where $d\tau = d^3\mathbf{x} dt$.

We know, however, in classical electrodynamics, that the Coulomb potential does not act instantaneously, but is delayed by a time r_{56}, taking the speed of light as unity. This suggests simply replacing $r_{56}^{-1}\delta(t_{56})$ in (2) by something like $r_{56}^{-1}\delta(t_{56} - r_{56})$ to represent the delay in the effect of b on a.

This turns out to be not quite right,[7] for when this interaction is represented by photons they must be of only positive energy, while the Fourier transform of $\delta(t_{56} - r_{56})$ contains frequencies of both signs. It should instead be replaced by $\delta_+(t_{56} - r_{56})$ where

$$\delta_+(x) = \int_0^\infty e^{-i\omega x}d\omega/\pi = \lim_{\epsilon \to 0} \frac{(\pi i)^{-1}}{x - i\epsilon} = \delta(x) + (\pi i x)^{-1}. \quad (3)$$

This is to be averaged with $r_{56}^{-1}\delta_+(-t_{56} - r_{56})$ which arises when $t_5 < t_6$ and corresponds to a emitting the quantum which b receives. Since

$$(2r)^{-1}(\delta_+(t-r) + \delta_+(-t-r)) = \delta_+(t^2 - r^2),$$

this means $r_{56}^{-1}\delta(t_{56})$ is replaced by $\delta_+(s_{56}^2)$ where $s_{56}^2 = t_{56}^2 - r_{56}^2$ is the square of the relativistically invariant interval between points 5 and 6. Since in classical electrodynamics there is also an interaction through the vector potential, the complete interaction (see A, Eq. (1)) should be $(1 - (\mathbf{v}_5 \cdot \mathbf{v}_6))\delta_+(s_{56}^2)$, or in the relativistic case,

$$(1 - \boldsymbol{\alpha}_a \cdot \boldsymbol{\alpha}_b)\delta_+(s_{56}^2) = \beta_a\beta_b\gamma_{a\mu}\gamma_{b\mu}\delta_+(s_{56}^2).$$

Hence we have for electrons obeying the Dirac equation,

$$K^{(1)}(3,4;1,2) = -ie^2 \iint K_{+a}(3,5)K_{+b}(4,6)\gamma_{a\mu}\gamma_{b\mu}$$
$$\times \delta_+(s_{56}^2)K_{+a}(5,1)K_{+b}(6,2)d\tau_5 d\tau_6, \quad (4)$$

where $\gamma_{a\mu}$ and $\gamma_{b\mu}$ are the Dirac matrices applying to the spinor corresponding to particles a and b, respectively (the factor $\beta_a\beta_b$ being absorbed in the definition, I Eq. (17), of K_+).

This is our fundamental equation for electrodynamics. It describes the effect of exchange of one quantum (therefore first order in e^2) between two electrons. It will serve as a prototype enabling us to write down the corresponding quantities involving the exchange of two or more quanta between two electrons or the interaction of an electron with itself. It is a consequence of conventional electrodynamics. Relativistic invariance is clear. Since one sums over μ it contains the effects of both longitudinal and transverse waves in a relativistically symmetrical way.

We shall now interpret Eq. (4) in a manner which will permit us to write down the higher order terms. It can be understood (see Fig. 1) as saying that the amplitude for "a" to go from 1 to 3 and "b" to go from 2 to 4 is altered to first order because they can exchange a quantum. Thus, "a" can go to 5 (amplitude $K_+(5,1)$)

---
[7] It, and a like term for the effect of a on b, leads to a theory which, in the classical limit, exhibits interaction through half-advanced and half-retarded potentials. Classically, this is equivalent to purely retarded effects within a closed box from which no light escapes (e.g., see A, or J. A. Wheeler and R. P. Feynman, Rev. Mod. Phys. 17, 157 (1945)). Analogous theorems exist in quantum mechanics but it would lead us too far astray to discuss them now.

emit a quantum (longitudinal, transverse, or scalar $\gamma_{a\mu}$) and then proceed to 3 ($K_+(3, 5)$). Meantime "b" goes to 6 ($K_+(6, 2)$), absorbs the quantum ($\gamma_{b\mu}$) and proceeds to 4 ($K_+(4, 6)$). The quantum meanwhile proceeds from 5 to 6, which it does with amplitude $\delta_+(s_{56}^2)$. We must sum over all the possible quantum polarizations μ and positions and times of emission 5, and of absorption 6. Actually if $t_5 > t_6$ it would be better to say that "a" absorbs and "b" emits but no attention need be paid to these matters, as all such alternatives are automatically contained in (4).

The correct terms of higher order in e^2 or involving larger numbers of electrons (interacting with themselves or in pairs) can be written down by the same kind of reasoning. They will be illustrated by examples as we proceed. In a succeeding paper they will all be deduced from conventional quantum electrodynamics.

Calculation, from (4), of the transition element between positive energy free electron states gives the Möller scattering of two electrons, when account is taken of the Pauli principle.

The exclusion principle for interacting charges is handled in exactly the same way as for non-interacting charges (I). For example, for two charges it requires only that one calculate $K(3, 4; 1, 2) - K(4, 3; 1, 2)$ to get the net amplitude for arrival of charges at 3 and 4. It is disregarded in intermediate states. The interference effects for scattering of electrons by positrons discussed by Bhabha will be seen to result directly in this formulation. The formulas are interpreted to apply to positrons in the manner discussed in I.

As our primary concern will be for processes in which the quanta are virtual we shall not include here the detailed analysis of processes involving real quanta in initial or final state, and shall content ourselves by only stating the rules applying to them.[8] The result of the analysis is, as expected, that they can be included by the same line of reasoning as is used in discussing the virtual processes, provided the quantities are normalized in the usual manner to represent single quanta. For example, the amplitude that an electron in going from 1 to 2 absorbs a quantum whose vector potential, suitably normalized, is $c_\mu \exp(-ik\cdot x) = C_\mu(x)$ is just the expression (I, Eq. (13)) for scattering in a potential with A (3) replaced by C (3). Each quantum interacts only

[8] Although in the expressions stemming from (4) the quanta are virtual, this is not actually a theoretical limitation. One way to deduce the correct rules for real quanta from (4) is to note that in a closed system all quanta can be considered as virtual (i.e., they have a known source and are eventually absorbed) so that in such a system the present description is complete and equivalent to the conventional one. In particular, the relation of the Einstein A and B coefficients can be deduced. A more practical direct deduction of the expressions for real quanta will be given in the subsequent paper. It may be noted that (4) can be rewritten as describing the action on a, $K^{(1)}(3, 1) = i \int K_+(3, 5) \times A(5) K_+(5, 1) d\tau_5$ of the potential $A_\mu(5) = e^2 \int K_+(4, 6) \delta_+(s_{56}^2) \gamma_\mu \times K_+(6, 2) d\tau_6$ arising from Maxwell's equations $-\Box^2 A_\mu = 4\pi j_\mu$ from a "current" $j_\mu(6) = e^2 K_+(4, 6) \gamma_\mu K_+(6, 2)$ produced by particle b in going from 2 to 4. This is virtue of the fact that δ_+ satisfies
$$-\Box_2^2 \delta_+(s_{21}^2) = 4\pi\delta(2, 1). \quad (5)$$

once (either in emission or in absorption), terms like (I, Eq. (14)) occur only when there is more than one quantum involved. The Bose statistics of the quanta can, in all cases, be disregarded in intermediate states. The only effect of the statistics is to change the weight of initial or final states. If there are among quanta, in the initial state, some n which are identical then the weight of the state is $(1/n!)$ of what it would be if these quanta were considered as different (similarly for the final state).

3. THE SELF-ENERGY PROBLEM

Having a term representing the mutual interaction of a pair of charges, we must include similar terms to represent the interaction of a charge with itself. For under some circumstances what appears to be two distinct electrons may, according to I, be viewed also as a single electron (namely in case one electron was created in a pair with a positron destined to annihilate the other electron). Thus to the interaction between such electrons must correspond the possibility of the action of an electron on itself.[9]

This interaction is the heart of the self energy problem. Consider to first order in e^2 the action of an electron on itself in an otherwise force free region. The amplitude $K(2, 1)$ for a single particle to get from 1 to 2 differs from $K_+(2, 1)$ to first order in e^2 by a term

$$K^{(1)}(2, 1) = -ie^2 \int\int K_+(2, 4) \gamma_\mu K_+(4, 3) \gamma_\mu$$
$$\times K_+(3, 1) d\tau_3 d\tau_4 \delta_+(s_{43}^2). \quad (6)$$

It arises because the electron instead of going from 1 directly to 2, may go (Fig. 2) first to 3, ($K_+(3, 1)$), emit a quantum (γ_μ), proceed to 4, ($K_+(4, 3)$), absorb it (γ_μ), and finally arrive at 2 ($K_+(2, 4)$). The quantum must go from 3 to 4 ($\delta_+(s_{43}^2)$).

This is related to the self-energy of a free electron in the following manner. Suppose initially, time t_1, we have an electron in state $f(1)$ which we imagine to be a positive energy solution of Dirac's equation for a free particle. After a long time $t_2 - t_1$ the perturbation will alter

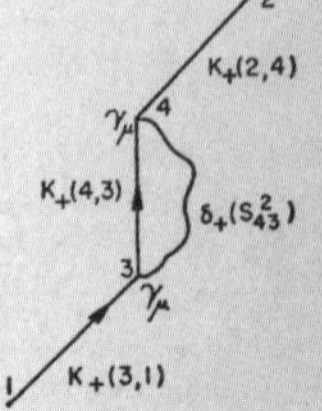

FIG. 2. Interaction of an electron with itself, Eq. (6).

[9] These considerations make it appear unlikely that the contention of J. A. Wheeler and R. P. Feynman, Rev. Mod. Phys. 17, 157 (1945), that electrons do not act on themselves, will be a successful concept in quantum electrodynamics.

the wave function, which can then be looked upon as a superposition of free particle solutions (actually it only contains f). The amplitude that $g(2)$ is contained is calculated as in (I, Eq. (21)). The diagonal element ($g=f$) is therefore

$$\int\int \bar{f}(2)\beta K^{(1)}(2,1)\beta f(1)d^3\mathbf{x}_1 d^3\mathbf{x}_2. \quad (7)$$

The time interval $T = t_2 - t_1$ (and the spatial volume V over which one integrates) must be taken very large, for the expressions are only approximate (analogous to the situation for two interacting charges).[10] This is because, for example, we are dealing incorrectly with quanta emitted just before t_2 which would normally be reabsorbed at times after t_2.

If $K^{(1)}(2,1)$ from (6) is actually substituted into (7) the surface integrals can be performed as was done in obtaining I, Eq. (22) resulting in

$$-ie^2\int\int \bar{f}(4)\gamma_\mu K_+(4,3)\gamma_\mu f(3)\delta_+(s_{43}{}^2)d\tau_3 d\tau_4. \quad (8)$$

Putting for $f(1)$ the plane wave $u \exp(-ip \cdot x_1)$ where p_μ is the energy (p_4) and momentum of the electron ($p^2 = m^2$), and u is a constant 4-index symbol, (8) becomes

$$-ie^2\int\int (\bar{u}\gamma_\mu K_+(4,3)\gamma_\mu u)$$
$$\times \exp(ip \cdot (x_4 - x_3))\delta_+(s_{43}{}^2)d\tau_3 d\tau_4,$$

the integrals extending over the volume V and time interval T. Since $K_+(4,3)$ depends only on the difference of the coordinates of 4 and 3, $x_{43\mu}$, the integral on 4 gives a result (except near the surfaces of the region) independent of 3. When integrated on 3, therefore, the result is of order VT. The effect is proportional to V, for the wave functions have been normalized to unit

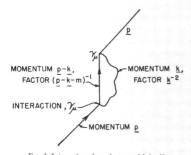

FIG. 3. Interaction of an electron with itself. Momentum space, Eq. (11).

---

[10] This is discussed in reference 5 in which it is pointed out that the concept of a wave function loses accuracy if there are delayed self-actions.

volume. If normalized to volume V, the result would simply be proportional to T. This is expected, for if the effect were equivalent to a change in energy ΔE, the amplitude for arrival in f at t_2 is altered by a factor $\exp(-i\Delta E(t_2 - t_1))$, or to first order by the difference $-i(\Delta E)T$. Hence, we have

$$\Delta E = e^2\int (\bar{u}\gamma_\mu K_+(4,3)\gamma_\mu u) \exp(ip \cdot x_{43})\delta_+(s_{43}{}^2)d\tau_4, \quad (9)$$

integrated over all space-time $d\tau_4$. This expression will be simplified presently. In interpreting (9) we have tacitly assumed that the wave functions are normalized so that $(u^*u) = (\bar{u}\gamma_4 u) = 1$. The equation may therefore be made independent of the normalization by writing the left side as $(\Delta E)(\bar{u}\gamma_4 u)$, or since $(\bar{u}\gamma_4 u) = (E/m)(\bar{u}u)$ and $m\Delta m = E\Delta E$, as $\Delta m(\bar{u}u)$ where Δm is an equivalent change in mass of the electron. In this form invariance is obvious.

One can likewise obtain an expression for the energy shift for an electron in a hydrogen atom. Simply replace K_+ in (8), by $K_+^{(V)}$, the exact kernel for an electron in the potential, $V = \beta e^2/r$, of the atom, and f by a wave function (of space and time) for an atomic state. In general the ΔE which results is not real. The imaginary part is negative and in $\exp(-i\Delta ET)$ produces an exponentially decreasing amplitude with time. This is because we are asking for the amplitude that an atom initially with no photon in the field, will still appear after time T with no photon. If the atom is in a state which can radiate, this amplitude must decay with time. The imaginary part of ΔE when calculated does indeed give the correct rate of radiation from atomic states. It is zero for the ground state and for a free electron.

In the non-relativistic region the expression for ΔE can be worked out as has been done by Bethe.[11] In the relativistic region (points 4 and 3 as close together as a Compton wave-length) the $K_+^{(V)}$ which should appear in (8) can be replaced to first order in V by K_+ plus $K_+^{(1)}(2,1)$ given in I, Eq. (13). The problem is then very similar to the radiationless scattering problem discussed below.

4. EXPRESSION IN MOMENTUM AND ENERGY SPACE

The evaluation of (9), as well as all the other more complicated expressions arising in these problems, is very much simplified by working in the momentum and energy variables, rather than space and time. For this we shall need the Fourier Transform of $\delta_+(s_{21}{}^2)$ which is

$$-\delta_+(s_{21}{}^2) = \pi^{-1}\int \exp(-ik \cdot x_{21})k^{-2}d^4k, \quad (10)$$

which can be obtained from (3) and (5) or from I, Eq. (32) noting that $I_+(2,1)$ for $m^2 = 0$ is $\delta_+(s_{21}{}^2)$ from

---

[11] H. A. Bethe, Phys. Rev. **72**, 339 (1947).

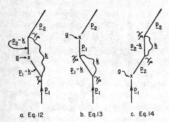
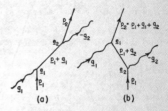

a. Eq.12 b. Eq.13 c. Eq.14

FIG. 4. Radiative correction to scattering, momentum space.

FIG. 5. Compton scattering, Eq. (15).

I, Eq. (34). The k^{-2} means $(k \cdot k)^{-1}$ or more precisely the limit as $\delta \to 0$ of $(k \cdot k + i\delta)^{-1}$. Further d^4k means $(2\pi)^{-2} dk_1 dk_2 dk_3 dk_4$. If we imagine that quanta are particles of zero mass, then we can make the general rule that all poles are to be resolved by considering the masses of the particles and quanta to have infinitesimal negative imaginary parts.

Using these results we see that the self-energy (9) is the matrix element between $\bar{u}$ and u of the matrix

$$(e^2/\pi i)\int \gamma_\mu (p-k-m)^{-1} \gamma_\mu k^{-2} d^4k, \quad (11)$$

where we have used the expression (I, Eq. (31)) for the Fourier transform of K_+. This form for the self-energy is easier to work with than is (9).

The equation can be understood by imagining (Fig. 3) that the electron of momentum p emits (γ_μ) a quantum of momentum k, and makes its way now with momentum $p-k$ to the next event (factor $(p-k-m)^{-1}$) which is to absorb the quantum (another γ_μ). The amplitude of propagation of quanta is k^{-2}. (There is a factor $e^2/\pi i$ for each virtual quantum.) One integrates over all quanta. The reason an electron of momentum p propagates as $1/(p-m)$ is that this operator is the reciprocal of the Dirac equation operator, and we are simply solving this equation. Likewise light goes as $1/k^2$, for this is the reciprocal D'Alembertian operator of the wave equation of light. The first γ_μ represents the current which generates the vector potential, while the second is the velocity operator by which this potential is multiplied in the Dirac equation when an external field acts on an electron.

Using the same line of reasoning, other problems may be set up directly in momentum space. For example, consider the scattering in a potential $A = A_\mu \gamma_\mu$ varying in space and time as $a \exp(-iq \cdot x)$. An electron initially in state of momentum $p_1 = p_{1\mu} \gamma_\mu$ will be deflected to state p_2 where $p_2 = p_1 + q$. The zero-order answer is simply the matrix element of a between states 1 and 2. We next ask for the first order (in e^2) radiative correction due to virtual radiation of one quantum. There are several ways this can happen. First for the case illus-

trated in Fig. 4(a), find the matrix:

$$(e^2/\pi i)\int \gamma_\mu (p_2-k-m)^{-1} a(p_1-k-m)^{-1} \gamma_\mu k^{-2} d^4k. \quad (12)$$

For in this case, first[12] a quantum of momentum k is emitted (γ_μ), the electron then having momentum p_1-k and hence propagating with factor $(p_1-k-m)^{-1}$. Next it is scattered by the potential (matrix a) receiving additional momentum q, propagating on then (factor $(p_2-k-m)^{-1}$) with the new momentum until the quantum is reabsorbed (γ_μ). The quantum propagates from emission to absorption (k^{-2}) and we integrate over all quanta (d^4k), and sum on polarization μ. When this is integrated on k_4, the result can be shown to be exactly equal to the expressions (16) and (17) given in B for the same process, the various terms coming from residues of the poles of the integrand (12).

Or again if the quantum is both emitted and reabsorbed before the scattering takes place one finds (Fig. 4(b))

$$(e^2/\pi i)\int a(p_1-m)^{-1} \gamma_\mu (p_1-k-m)^{-1} \gamma_\mu k^{-2} d^4k, \quad (13)$$

or if both emission and absorption occur after the scattering, (Fig. 4(c))

$$(e^2/\pi i)\int \gamma_\mu (p_2-k-m)^{-1} \gamma_\mu (p_2-m)^{-1} a k^{-2} d^4k. \quad (14)$$

These terms are discussed in detail below.

We have now achieved our simplification of the form of writing matrix elements arising from virtual processes. Processes in which a number of real quanta is given initially and finally offer no problem (assuming correct normalization). For example, consider the Compton effect (Fig. 5(a)) in which an electron in state p_1 absorbs a quantum of momentum q_1, polarization vector $e_{1\mu}$ so that its interaction is $e_{1\mu} \gamma_\mu = e_1$, and emits a second quantum of momentum $-q_2$, polarization e_2 to arrive in final state of momentum p_2. The matrix for

[12] First, next, etc., here refer not to the order in true time but to the succession of events along the trajectory of the electron. That is, more precisely, to the order of appearance of the matrices in the expressions.

this process is $e_2(p_1+q_1-m)^{-1}e_1$. The total matrix for the Compton effect is, then,

$$e_2(p_1+q_1-m)^{-1}e_1+e_1(p_1+q_2-m)^{-1}e_2, \quad (15)$$

the second term arising because the emission of e_2 may also precede the absorption of e_1 (Fig. 5(b)). One takes matrix elements of this between initial and final electron states ($p_1+q_1=p_2-q_2$), to obtain the Klein Nishina formula. Pair annihilation with emission of two quanta, etc., are given by the same matrix, positron states being those with negative time component of p. Whether quanta are absorbed or emitted depends on whether the time component of q is positive or negative.

5. THE CONVERGENCE OF PROCESSES WITH VIRTUAL QUANTA

These expressions are, as has been indicated, no more than a re-expression of conventional quantum electrodynamics. As a consequence, many of them are meaningless. For example, the self-energy expression (9) or (11) gives an infinite result when evaluated. The infinity arises, apparently, from the coincidence of the δ-function singularities in $K_+(4, 3)$ and $\delta_+(s_{43}^2)$. Only at this point is it necessary to make a real departure from conventional electrodynamics, a departure other than simply rewriting expressions in a simpler form.

We desire to make a modification of quantum electrodynamics analogous to the modification of classical electrodynamics described in a previous article, A. There the $\delta(s_{12}^2)$ appearing in the action of interaction was replaced by $f(s_{12}^2)$ where $f(x)$ is a function of small width and great height.

The obvious corresponding modification in the quantum theory is to replace the $\delta_+(s^2)$ appearing the quantum mechanical interaction by a new function $f_+(s^2)$. We can postulate that if the Fourier transform of the classical $f(s_{12}^2)$ is the integral over all k of $F(k^2)\exp(-ik\cdot x_{12})d^4k$, then the Fourier transform of $f_+(s^2)$ is the same integral taken over only positive frequencies k_4 for $l_2>l_1$ and over only negative ones for $l_2<l_1$ in analogy to the relation of $\delta_+(s^2)$ to $\delta(s^2)$. The function $f(s^2)=f(x\cdot x)$ can be written* as

$$f(x\cdot x)=(2\pi)^{-2}\int_{k_4=0}^{\infty}\int \sin(k_4|x_4|)$$
$$\times \cos(\mathbf{K}\cdot\mathbf{x})dk_4 d^3\mathbf{K} g(k\cdot k),$$

where $g(k\cdot k)$ is k_4^{-1} times the density of oscillators and may be expressed for positive k_4 as (A, Eq. (16))

$$g(k^2)=\int_0^{\infty}(\delta(k^2)-\delta(k^2-\lambda^2))G(\lambda)d\lambda,$$

where $\int_0^{\infty}G(\lambda)d\lambda=1$ and G involves values of λ large compared to m. This simply means that the amplitude

* This relation is given incorrectly in A, equation just preceding 16.

for propagation of quanta of momentum k is

$$-F_+(k^2)=\pi^{-1}\int_0^{\infty}(k^{-2}-(k^2-\lambda^2)^{-1})G(\lambda)d\lambda,$$

rather than k^{-2}. That is, writing $F_+(k^2)=-\pi^{-1}k^{-2}C(k^2)$,

$$-f_+(s_{12}^2)=\pi^{-1}\int \exp(-ik\cdot x_{12})k^{-2}C(k^2)d^4k. \quad (16)$$

Every integral over an intermediate quantum which previously involved a factor d^4k/k^2 is now supplied with a convergence factor $C(k^2)$ where

$$C(k^2)=\int_0^{\infty}-\lambda^2(k^2-\lambda^2)^{-1}G(\lambda)d\lambda. \quad (17)$$

The poles are defined by replacing k^2 by $k^2+i\delta$ in the limit $\delta\to 0$. That is λ^2 may be assumed to have an infinitesimal negative imaginary part.

The function $f_+(s_{12}^2)$ may still have a discontinuity in value on the light cone. This is of no influence for the Dirac electron. For a particle satisfying the Klein Gordon equation, however, the interaction involves gradients of the potential which reinstates the δ function if f has discontinuities. The condition that f is to have no discontinuity in value on the light cone implies $k^2C(k^2)$ approaches zero as k^2 approaches infinity. In terms of $G(\lambda)$ the condition is

$$\int_0^{\infty}\lambda^2 G(\lambda)d\lambda=0. \quad (18)$$

This condition will also be used in discussing the convergence of vacuum polarization integrals.

The expression for the self-energy matrix is now

$$(e^2/\pi i)\int \gamma_\mu(p-k-m)^{-1}\gamma_\mu k^{-2}d^4kC(k^2), \quad (19)$$

which, since $C(k^2)$ falls off at least as rapidly as $1/k^2$, converges. For practical purposes we shall suppose hereafter that $C(k^2)$ is simply $-\lambda^2/(k^2-\lambda^2)$ implying that some average (with weight $G(\lambda)d\lambda$) over values of λ may be taken afterwards. Since in all processes the quantum momentum will be contained in at least one extra factor of the form $(p-k-m)^{-1}$ representing propagation of an electron while that quantum is in the field, we can expect all such integrals with their convergence factors to converge and that the result of all such processes will now be finite and definite (excepting the processes with closed loops, discussed below, in which the diverging integrals are over the momenta of the electrons rather than the quanta).

The integral of (19) with $C(k^2)=-\lambda^2(k^2-\lambda^2)^{-1}$ noting that $p^2=m^2$, $\lambda\gg m$ and dropping terms of order m/λ, is (see Appendix A)

$$(e^2/2\pi)[4m(\ln(\lambda/m)+\tfrac{1}{2})-p(\ln(\lambda/m)+5/4)]. \quad (20)$$

When applied to a state of an electron of momentum p satisfying $pu=mu$, it gives for the change in mass (as in B, Eq. (9))

$$\Delta m = m(e^2/2\pi)(3\ln(\lambda/m)+\tfrac{3}{4}). \qquad (21)$$

6. RADIATIVE CORRECTIONS TO SCATTERING

We can now complete the discussion of the radiative corrections to scattering. In the integrals we include the convergence factor $C(k^2)$, so that they converge for large k. Integral (12) is also not convergent because of the well-known infra-red catastrophy. For this reason we calculate (as discussed in B) the value of the integral assuming the photons to have a small mass $\lambda_{\min} \ll m \ll \lambda$. The integral (12) becomes

$$(e^2/\pi i)\int \gamma_\mu (p_2-k-m)^{-1} a(p_1-k-m)^{-1}$$

$$\times \gamma_\mu (k^2-\lambda_{\min}{}^2)^{-1} d^4k C(k^2-\lambda_{\min}{}^2),$$

which when integrated (see Appendix B) gives $(e^2/2\pi)$ times

$$\left[2\left(\ln\frac{m}{\lambda_{\min}}-1\right)\left(1-\frac{2\theta}{\tan 2\theta}\right)+\theta\tan\theta \right.$$

$$+\frac{4}{\tan 2\theta}\int_0^\theta \alpha\tan\alpha\, d\alpha \Big]a$$

$$+\frac{1}{4m}(qa-aq)\frac{2\theta}{\sin 2\theta}+ra, \quad (22)$$

where $(q^2)^{\frac{1}{2}} = 2m\sin\theta$ and we have assumed the matrix to operate between states of momentum p_1 and $p_2=p_1+q$ and have neglected terms of order $\lambda_{\min}/m$, m/λ, and q^2/λ^2. Here the only dependence on the convergence factor is in the term ra, where

$$r=\ln(\lambda/m)+9/4-2\ln(m/\lambda_{\min}). \qquad (23)$$

As we shall see in a moment, the other terms (13), (14) give contributions which just cancel the ra term. The remaining terms give for small q,

$$(e^2/4\pi)\left(\frac{1}{2m}(qa-aq)+\frac{4q^2}{3m^2}a\left(\ln\frac{m}{\lambda_{\min}}-\frac{3}{8}\right)\right), \quad (24)$$

which shows the change in magnetic moment and the Lamb shift as interpreted in more detail in B.[13]

[13] That the result given in B in Eq. (19) was in error was repeatedly pointed out to the author, in private communication, by V. F. Weisskopf and J. B. French, as their calculation, completed simultaneously with the author's early in 1948, gave a different result. French has finally shown that although the expression for the radiationless scattering B, Eq. (18) or (24) above is correct, it was incorrectly joined onto Bethe's non-relativistic result. He shows that the relation $\ln 2k_{\max} -1 = \ln\lambda_{\min}$ used by the author should have been $\ln 2k_{\max} - 5/6 = \ln\lambda_{\min}$. This results in adding a term $-(1/6)$ to the logarithm in B, Eq. (19) so that the result now agrees with that of J. B. French and V. F. Weisskopf,

We must now study the remaining terms (13) and (14). The integral on k in (13) can be performed (after multiplication by $C(k^2)$) since it involves nothing but the integral (19) for the self-energy and the result is allowed to operate on the initial state u_1, (so that $p_1 u_1 = m u_1$). Hence the factor following $a(p_1-m)^{-1}$ will be just Δm. But, if one now tries to expand $1/(p_1-m) = (p_1+m)/(p_1^2-m^2)$ one obtains an infinite result, since $p_1^2=m^2$. This is, however, just what is expected physically. For the quantum can be emitted and absorbed at any time previous to the scattering. Such a process has the effect of a change in mass of the electron in the state 1. It therefore changes the energy by ΔE and the amplitude to first order in ΔE by $-i\Delta E \cdot t$ where t is the time it is acting, which is infinite. That is, the major effect of this term would be canceled by the effect of change of mass Δm.

The situation can be analyzed in the following manner. We suppose that the electron approaching the scattering potential a has not been free for an infinite time, but at some time far past suffered a scattering by a potential b. If we limit our discussion to the effects of Δm and of the virtual radiation of one quantum between two such scatterings each of the effects will be finite, though large, and their difference is determinate. The propagation from b to a is represented by a matrix

$$a(p'-m)^{-1}b, \qquad (25)$$

in which one is to integrate possibly over p' (depending on details of the situation). (If the time is long between b and a, the energy is very nearly determined so that p'^2 is very nearly m^2.)

We shall compare the effect on the matrix (25) of the virtual quanta and of the change of mass Δm. The effect of a virtual quantum is

$$(e^2/\pi i)\int a(p'-m)^{-1}\gamma_\mu(p'-k-m)^{-1}$$

$$\times \gamma_\mu(p'-m)^{-1} b k^{-2} d^4k C(k^2), \quad (26)$$

while that of a change of mass can be written

$$a(p'-m)^{-1}\Delta m(p'-m)^{-1}b, \qquad (27)$$

and we are interested in the difference (26)-(27). A simple and direct method of making this comparison is just to evaluate the integral on k in (26) and subtract from the result the expression (27) where Δm is given in (21). The remainder can be expressed as a multiple $-r(p'^2)$ of the unperturbed amplitude (25);

$$-r(p'^2)a(p'-m)^{-1}b. \qquad (28)$$

This has the same result (to this order) as replacing the potentials a and b in (25) by $(1-\tfrac{1}{2}r(p'^2))a$ and

Phys. Rev. **75**, 1240 (1949) and N. H. Kroll and W. E. Lamb, Phys. Rev. **75**, 388 (1949). The author feels unhappily responsible for the very considerable delay in the publication of French's result occasioned by this error. This footnote is appropriately numbered.

$(1-\frac{1}{2}r(p'^2))b$. In the limit, then, as $p'^2 \to m^2$ the net effect on the scattering is $-\frac{1}{2}ra$ where r, the limit of $r(p'^2)$ as $p'^2 \to m^2$ (assuming the integrals have an infrared cut-off), turns out to be just equal to that given in (23). An equal term $-\frac{1}{2}ra$ arises from virtual transitions after the scattering (14) so that the entire ra term in (22) is canceled.

The reason that r is just the value of (12) when $q^2 = 0$ can also be seen without a direct calculation as follows: Let us call p the vector of length m in the direction of p' so that if $p'^2 = m(1+\epsilon)^2$ we have $p' = (1+\epsilon)p$ and we take ϵ as very small, being of order T^{-1} where T is the time between the scatterings b and a. Since $(p'-m)^{-1} = (p'+m)/(p'^2-m^2) \approx (p+m)/2m^2\epsilon$, the quantity (25) is of order ϵ^{-1} or T. We shall compute corrections to it only to its own order (ϵ^{-1}) in the limit $\epsilon \to 0$. The term (27) can be written approximately[14] as

$$(e^2/\pi i) \int a(p'-m)^{-1}\gamma_\mu(p-k-m)^{-1}$$
$$\times \gamma_\mu(p'-m)^{-1} bk^{-2}d^4kC(k^2),$$

using the expression (19) for Δm. The net of the two effects is therefore approximately[15]

$$-(e^2/\pi i) \int a(p'-m)^{-1}\gamma_\mu(p-k-m)^{-1}\epsilon p(p-k-m)^{-1}$$
$$\times \gamma_\mu(p'-m)^{-1} bk^{-2}d^4kC(k^2),$$

a term now of order $1/\epsilon$ (since $(p'-m)^{-1} \approx (p+m)$ $\times (2m^2\epsilon)^{-1}$) and therefore the one desired in the limit. Comparison to (28) gives for r the expression

$$(p_1+m/2m) \int \gamma_\mu(p_1-k-m)^{-1}(p_1m^{-1})(p_1-k-m)^{-1}$$
$$\times \gamma_\mu k^{-2}d^4kC(k^2). \quad (29)$$

The integral can be immediately evaluated, since it is the same as the integral (12), but with $q = 0$, for a replaced by p_1/m. The result is therefore $r \cdot (p_1/m)$ which when acting on the state u_1 is just r, as $p_1u_1 = mu_1$. For the same reason the term $(p_1+m)/2m$ in (29) is effectively 1 and we are left with $-r$ of (23).[16]

In more complex problems starting with a free elec-

---

[14] The expression is not exact because the substitution of Δm by the integral in (19) is valid only if p operates on a state such that p can be replaced by m. The error, however, is of order $a(p'-m)^{-1}(p-m)(p'-m)^{-1}b$ which is $a((1+\epsilon)p+m)(p-m)$ $\times((1+\epsilon)p+m)p(2\epsilon+\epsilon^2)^{-1}m^{-4}$. But since $p^2 = m^2$, we have $p(p-m) = -m(p-m) = (p-m)p$ so the net result is approximately $a(p-m)b/4m^2$ and is not of order $1/\epsilon$ but smaller, so that its effect drops out in the limit.

[15] We have used, to first order, the general expansion (valid for any operators A, B)
$$(A+B)^{-1} = A^{-1} - A^{-1}BA^{-1} + A^{-1}BA^{-1}BA^{-1} - \cdots$$
with $A = p-k-m$ and $B = p'-p = \epsilon p$ to expand the difference of $(p'-k-m)^{-1}$ and $(p-k-m)^{-1}$.

[16] The renormalization terms appearing B, Eqs. (14), (15) when translated directly into the present notation do not give twice (29) but give this expression with the central p_1m^{-1} factor replaced by $m\gamma_4/E_1$ where $E_1 = p_{1\mu}$ for $\mu = 4$. When integrated it therefore gives $ra((p_1+m)/2m)(m\gamma_4/E_1)$ or $ra - ra(m\gamma_4/E_1)(p_1-m)/2m$. (Since $p_1\gamma_4 + \gamma_4 p_1 = 2E_1$) which gives just ra, since $p_1u_1 = mu_1$.

---

tron the same type of term arises from the effects of a virtual emission and absorption both previous to the other processes. They, therefore, simply lead to the same factor r so that the expression (23) may be used directly and these renormalization integrals need not be computed afresh for each problem.

In this problem of the radiative corrections to scattering the net result is insensitive to the cut-off. This means, of course, that by a simple rearrangement of terms previous to the integration we could have avoided the use of the convergence factors completely (see for example Lewis[17]). The problem was solved in the manner here in order to illustrate how the use of such convergence factors, even when they are actually unnecessary, may facilitate analysis somewhat by removing the effort and ambiguities that may be involved in trying to rearrange the otherwise divergent terms.

The replacement of δ_+ by f_+ given in (16), (17) is not determined by the analogy with the classical problem. In the classical limit only the real part of δ_+ (i.e., just δ) is easy to interpret. But by what should the imaginary part, $1/(\pi i s^2)$, of δ_+ be replaced? The choice we have made here (in defining, as we have, the location of the poles of (17)) is arbitrary and almost certainly incorrect. If the radiation resistance is calculated for an atom, as the imaginary part of (8), the result depends slightly on the function f_+. On the other hand the light radiated at very large distances from a source is independent of f_+. The total energy absorbed by distant absorbers will not check with the energy loss of the source. We are in a situation analogous to that in the classical theory if the entire f function is made to contain only retarded contributions (see A, Appendix). One desires instead the analogue of $\langle F \rangle_{\text{ret}}$ of A. This problem is being studied.

One can say therefore, that this attempt to find a consistent modification of quantum electrodynamics is incomplete (see also the question of closed loops, below). For it could turn out that any correct form of f_+ which will guarantee energy conservation may at the same time not be able to make the self-energy integral finite. The desire to make the methods of simplifying the calculation of quantum electrodynamic processes more widely available has prompted this publication before an analysis of the correct form for f_+ is complete. One might try to take the position that, since the energy discrepancies discussed vanish in the limit $\lambda \to \infty$, the correct physics might be considered to be that obtained by letting $\lambda \to \infty$ after mass renormalization. I have no proof of the mathematical consistency of this procedure, but the presumption is very strong that it is satisfactory. (It is also strong that a satisfactory form for f_+ can be found.)

7. THE PROBLEM OF VACUUM POLARIZATION

In the analysis of the radiative corrections to scattering one type of term was not considered. The potential

---

[17] H. W. Lewis, Phys. Rev. 73, 173 (1948).

which we can assume to vary as $a_\mu \exp(-iq \cdot x)$ creates a pair of electrons (see Fig. 6), momenta $p_a, -p_b$. This pair then reannihilates, emitting a quantum $q = p_b - p_a$, which quantum scatters the original electron from state 1 to state 2. The matrix element for this process (and the others which can be obtained by rearranging the order in time of the various events) is

FIG. 6. Vacuum polarization effect on scattering, Eq. (30).

$$-(e^2/\pi i)(\bar{u}_2 \gamma_\mu u_1) \int Sp[(p_a + q - m)^{-1}$$
$$\times \gamma_\nu (p_a - m)^{-1} \gamma_\mu] d^4 p_a q^{-2} C(q^2) a_\nu. \quad (30)$$

This is because the potential produces the pair with amplitude proportional to $a_\nu \gamma_\nu$, the electrons of momenta p_a and $-(p_a + q)$ proceed from there to annihilate, producing a quantum (factor γ_μ) which propagates (factor $q^{-2} C(q^2)$) over to the other electron, by which it is absorbed (matrix element of γ_μ between states 1 and 2 of the original electron ($\bar{u}_2 \gamma_\mu u_1$)). All momenta p_a and spin states of the virtual electron are admitted, which means the spur and the integral on $d^4 p_a$ are calculated.

One can imagine that the closed loop path of the positron-electron produces a current

$$4\pi j_\mu = J_{\mu\nu} a_\nu, \quad (31)$$

which is the source of the quanta which act on the second electron. The quantity

$$J_{\mu\nu} = -(e^2/\pi i) \int Sp[(p + q - m)^{-1}$$
$$\times \gamma_\nu (p - m)^{-1} \gamma_\mu] d^4 p, \quad (32)$$

is then characteristic for this problem of polarization of the vacuum.

One sees at once that $J_{\mu\nu}$ diverges badly. The modification of δ to f alters the amplitude with which the current j_μ will affect the scattered electron, but it can do nothing to prevent the divergence of the integral (32) and of its effects.

One way to avoid such difficulties is apparent. From one point of view we are considering all routes by which a given electron can get from one region of space-time to another, i.e., from the source of electrons to the apparatus which measures them. From this point of view the closed loop path leading to (32) is unnatural. It might be assumed that the only paths of meaning are those which start from the source and work their way in a continuous path (possibly containing many time reversals) to the detector. Closed loops would be excluded. We have already found that this may be done for electrons moving in a fixed potential.

Such a suggestion must meet several questions, however. The closed loops are a consequence of the usual hole theory in electrodynamics. Among other things, they are required to keep probability conserved. The probability that no pair is produced by a potential is not unity and its deviation from unity arises from the imaginary part of $J_{\mu\nu}$. Again, with closed loops excluded, a pair of electrons once created cannot annihilate one another again, the scattering of light by light would be zero, etc. Although we are not experimentally sure of these phenomena, this does seem to indicate that the closed loops are necessary. To be sure, it is always possible that these matters of probability conservation, etc., will work themselves out as simply in the case of interacting particles as for those in a fixed potential. Lacking such a demonstration the presumption is that the difficulties of vacuum polarization are not so easily circumvented.[18]

An alternative procedure discussed in B is to assume that the function $K_+(2,1)$ used above is incorrect and is to be replaced by a modified function K_+' having no singularity on the light cone. The effect of this is to provide a convergence factor $C(p^2 - m^2)$ for every integral over electron momenta.[19] This will multiply the integrand of (32) by $C(p^2 - m^2) C((p+q)^2 - m^2)$, since the integral was originally $\delta(p_a - p_b + q) d^4 p_a d^4 p_b$ and both p_a and p_b get convergence factors. The integral now converges but the result is unsatisfactory.[20]

One expects the current (31) to be conserved, that is $q_\mu j_\mu = 0$ or $q_\mu J_{\mu\nu} = 0$. Also one expects no current if a_ν is a gradient, or $a_\nu = q_\nu$ times a constant. This leads to the condition $J_{\mu\nu} q_\nu = 0$ which is equivalent to $q_\mu J_{\mu\nu} = 0$ since $J_{\mu\nu}$ is symmetrical. But when the expression (32) is integrated with such convergence factors it does not satisfy this condition. By altering the kernel from K to another, K', which does not satisfy the Dirac equation we have lost the gauge invariance, its consequent current conservation and the general consistency of the theory.

One can see this best by calculating $J_{\mu\nu} q_\nu$ directly from (32). The expression within the spur becomes $(p + q - m)^{-1} q (p - m)^{-1} \gamma_\mu$ which can be written as the difference of two terms: $(p - m)^{-1} \gamma_\mu - (p + q - m)^{-1} \gamma_\mu$. Each of these terms would give the same result if the integration $d^4 p$ were without a convergence factor, for

---

[18] It would be very interesting to calculate the Lamb shift accurately enough to be sure that the 20 megacycles expected from vacuum polarization are actually present.

[19] This technique also makes self-energy and radiationless scattering integrals finite even without the modification of δ_+ to f_+ for the radiation (and the consequent convergence factor $C(k^2)$ for the quanta). See B.

[20] Added to the terms given below (33) there is a term $\frac{1}{4}(\lambda^2 - 2\mu^2 + \frac{1}{2}q^2)\delta_{\mu\nu}$ for $C(k^2) = -\lambda^2(k^2 - \lambda^2)^{-1}$, which is not gauge invariant. (In addition the charge renormalization has $-7/6$ added to the logarithm.)

the first can be converted into the second by a shift of the origin of p, namely $p'=p+q$. This does not result in cancelation in (32) however, for the convergence factor is altered by the substitution.

A method of making (32) convergent without spoiling the gauge invariance has been found by Bethe and by Pauli. The convergence factor for light can be looked upon as the result of superposition of the effects of quanta of various masses (some contributing negatively). Likewise if we take the factor $C(p^2-m^2) = -\lambda^2(p^2-m^2-\lambda^2)^{-1}$ so that $(p^2-m^2)^{-1}C(p^2-m^2) = (p^2-m^2)^{-1} - (p^2-m^2-\lambda^2)^{-1}$ we are taking the difference of the result for electrons of mass m and mass $(\lambda^2+m^2)^{\frac{1}{2}}$. But we have taken this difference for *each* propagation between interactions with photons. They suggest instead that once created with a certain mass the electron should continue to propagate with this mass through all the potential interactions until it closes its loop. That is if the quantity (32), integrated over some finite range of p, is called $J_{\mu\nu}(m^2)$ and the corresponding quantity over the same range of p, but with m replaced by $(m^2+\lambda^2)^{\frac{1}{2}}$ is $J_{\mu\nu}(m^2+\lambda^2)$ we should calculate

$$J_{\mu\nu}{}^P = \int_0^\infty [J_{\mu\nu}(m^2) - J_{\mu\nu}(m^2+\lambda^2)]G(\lambda)d\lambda, \quad (32')$$

the function $G(\lambda)$ satisfying $\int_0^\infty G(\lambda)d\lambda = 1$ and $\int_0^\infty G(\lambda)\lambda^2 d\lambda = 0$. Then in the expression for $J_{\mu\nu}{}^P$ the range of p integration can be extended to infinity as the integral now converges. The result of the integration using this method is the integral on $d\lambda$ over $G(\lambda)$ of (see Appendix C)

$$J_{\mu\nu}{}^P = -\frac{e^2}{\pi}(q_\mu q_\nu - \delta_{\mu\nu}q^2)\left(\frac{1}{3}\ln\frac{\lambda^2}{m^2}\right.$$
$$\left. -\left[\frac{4m^2+2q^2}{3q^2}\left(1-\frac{\theta}{\tan\theta}\right)-\frac{1}{9}\right]\right), \quad (33)$$

with $q^2 = 4m^2 \sin^2\theta$.

The gauge invariance is clear, since $q_\mu(q_\mu q_\nu - q^2\delta_{\mu\nu}) = 0$. Operating (as it always will) on a potential of zero divergence the $(q_\mu q_\nu - \delta_{\mu\nu}q^2)a_\nu$ is simply $-q^2 a_\mu$, the D'Alembertian of the potential, that is, the current producing the potential. The term $-\frac{1}{3}(\ln(\lambda^2/m^2))(q_\mu q_\nu - q^2\delta_{\mu\nu})$ therefore gives a current proportional to the current producing the potential. This would have the same effect as a change in charge, so that we would have a difference $\Delta(e^2)$ between e^2 and the experimentally observed charge, $e^2 + \Delta(e^2)$, analogous to the difference between m and the observed mass. This charge depends logarithmically on the cut-off, $\Delta(e^2)/e^2 = -(2e^2/3\pi)\ln(\lambda/m)$. After this renormalization of charge is made, no effects will be sensitive to the cut-off.

After this is done the final term remaining in (33), contains the usual effects[21] of polarization of the vacuum.

[21] E. A. Uehling, Phys. Rev. **48**, 55 (1935); R. Serber, Phys. Rev. **48**, 49 (1935).

It is zero for a free light quantum ($q^2=0$). For small q^2 it behaves as $(2/15)q^2$ (adding $-\frac{1}{5}$ to the logarithm in the Lamb effect). For $q^2 > (2m)^2$ it is complex, the imaginary part representing the loss in amplitude required by the fact that the probability that no quanta are produced by a potential able to produce pairs $((q^2)^{\frac{1}{2}} > 2m)$ decreases with time. (To make the necessary analytic continuation, imagine m to have a small negative imaginary part, so that $(1-q^2/4m^2)^{\frac{1}{2}}$ becomes $-i(q^2/4m^2-1)^{\frac{1}{2}}$ as q^2 goes from below to above $4m^2$. Then $\theta = \pi/2 + iu$ where $\sinh u = +(q^2/4m^2-1)^{\frac{1}{2}}$, and $-1/\tan\theta = i\tanh u = +i(q^2-4m^2)^{\frac{1}{2}}(q^2)^{-\frac{1}{2}}$.)

Closed loops containing a number of quanta or potential interactions larger than two produce no trouble. Any loop with an odd number of interactions gives zero (I, reference 9). Four or more potential interactions give integrals which are convergent even without a convergence factor as is well known. The situation is analogous to that for self-energy. Once the simple problem of a single closed loop is solved there are no further divergence difficulties for more complex processes.[22]

8. LONGITUDINAL WAVES

In the usual form of quantum electrodynamics the longitudinal and transverse waves are given separate treatment. Alternately the condition $(\partial A_\mu/\partial x_\mu)\Psi = 0$ is carried along as a supplementary condition. In the present form no such special considerations are necessary for we are dealing with the solutions of the equation $-\Box^2 A_\mu = 4\pi j_\mu$ with a current j_μ which is conserved $\partial j_\mu/\partial x_\mu = 0$. That means at least $\Box^2(\partial A_\mu/\partial x_\mu) = 0$ and in fact our solution also satisfies $\partial A_\mu/\partial x_\mu = 0$.

To show that this is the case we consider the amplitude for emission (real or virtual) of a photon and show that the divergence of this amplitude vanishes. The amplitude for emission for photons polarized in the μ direction involves matrix elements of γ_μ. Therefore what we have to show is that the corresponding matrix elements of $q_\mu\gamma_\mu = q$ vanish. For example, for a first order effect we would require the matrix element of q between two states p_1 and $p_2 = p_1 + q$. But since $q = p_2 - p_1$ and $(\bar{u}_2 p_1 u_1) = m(\bar{u}_2 u_1) = (\bar{u}_2 p_2 u_1)$ the matrix element vanishes, which proves the contention in this case. It also vanishes in more complex situations (essentially because of relation (34), below) (for example, try putting $e_2 = q_2$ in the matrix (15) for the Compton Effect).

To prove this in general, suppose a_i, $i = 1$ to N are a set of plane wave disturbing potentials carrying momenta q_i (e.g., some may be emissions or absorptions of the same or different quanta) and consider a matrix for the transition from a state of momentum p_0 to p_N such

[22] There are loops completely without external interactions. For example, a pair is created virtually along with a photon. Next they annihilate, absorbing this photon. Such loops are disregarded on the grounds that they do not interact with anything and are thereby completely unobservable. Any indirect effects they may have via the exclusion principle have already been included.

as $a_N \prod_{i=1}^{N-1} (p_i-m)^{-1} a_i$ where $p_i = p_{i-1} + q_i$ (and in the product, terms with larger i are written to the left). The most general matrix element is simply a linear combination of these. Next consider the matrix between states p_0 and $p_N + q$ in a situation in which not only are the a_i acting but also another potential $a \exp(-iq \cdot x)$ where $a = q$. This may act previous to all a_i, in which case it gives $a_N \prod (p_i + q - m)^{-1} a_i (p_0 + q - m)^{-1} q$ which is equivalent to $+ a_N \prod (p_i + q - m)^{-1} a_i$ since $+(p_0 + q - m)^{-1} q$ is equivalent to $(p_0 + q - m)^{-1} \times (p_0 + q - m)$ as p_0 is equivalent to m acting on the initial state. Likewise if it acts after all the potentials it gives $q(p_N - m)^{-1} a_N \prod (p_i - m)^{-1} a_i$ which is equivalent to $-a_N \prod (p_i - m)^{-1} a_i$ since $p_N + q - m$ gives zero on the final state. Or again it may act between the potential a_k and a_{k+1} for each k. This gives

$$\sum_{k=1}^{N-1} a_N \prod_{i=k+1}^{N-1} (p_i + q - m)^{-1} a_i (p_k + q - m)^{-1}$$
$$\times q(p_k - m)^{-1} a_k \prod_{j=1}^{k-1} (p_j - m)^{-1} a_j.$$

However,

$$(p_k + q - m)^{-1} q (p_k - m)^{-1}$$
$$= (p_k - m)^{-1} - (p_k + q - m)^{-1}, \quad (34)$$

so that the sum breaks into the difference of two sums, the first of which may be converted to the other by the replacement of k by $k - 1$. There remain only the terms from the ends of the range of summation,

$$+ a_N \prod_{i=1}^{N-1} (p_i - m)^{-1} a_i - a_N \prod_{i=1}^{N-1} (p_i + q - m)^{-1} a_i.$$

These cancel the two terms originally discussed so that the entire effect is zero. Hence any wave emitted will satisfy $\partial A_\mu / \partial x_\mu = 0$. Likewise longitudinal waves (that is, waves for which $A_\mu = \partial \phi / \partial x_\mu$ or $a = q$) cannot be absorbed and will have no effect, for the matrix elements for emission and absorption are similar. (We have said little more than that a potential $A_\mu = \partial \varphi / \partial x_\mu$ has no effect on a Dirac electron since a transformation $\psi' = \exp(-i\phi)\psi$ removes it. It is also easy to see in coordinate representation using integrations by parts.)

This has a useful practical consequence in that in computing probabilities for transition for unpolarized light one can sum the squared matrix over all four directions rather than just the two special polarization vectors. Thus suppose the matrix element for some process for light polarized in direction e_μ is $e_\mu M_\mu$. If the light has wave vector q_μ we know from the argument above that $q_\mu M_\mu = 0$. For unpolarized light progressing in the z direction we would ordinarily calculate $M_x^2 + M_y^2$. But we can as well sum $M_x^2 + M_y^2 + M_z^2 - M_t^2$ for $q_\mu M_\mu$ implies $M_t = M_z$ since $q_t = q_z$ for free quanta. This shows that unpolarized light is a relativistically invariant concept, and permits some simplification in computing cross sections for such light.

Incidentally, the virtual quanta interact through terms like $\gamma_\mu \cdots \gamma_\mu k^{-2} d^4 k$. Real processes correspond to poles in the formulae for virtual processes. The pole occurs when $k^2 = 0$, but it looks at first as though in the sum on all four values of μ, of $\gamma_\mu \cdots \gamma_\mu$ we would have four kinds of polarization instead of two. Now it is clear that only two perpendicular to k are effective.

The usual elimination of longitudinal and scalar virtual photons (leading to an instantaneous Coulomb potential) can of course be performed here too (although it is not particularly useful). A typical term in a virtual transition is $\gamma_\mu \cdots \gamma_\mu k^{-2} d^4 k$ where the $\cdots$ represent some intervening matrices. Let us choose for the values of μ, the time t, the direction of vector part $\mathbf{K}$, of k, and two perpendicular directions 1, 2. We shall not change the expression for these two 1, 2 for these are represented by transverse quanta. But we must find $(\gamma_t \cdots \gamma_t) - (\gamma_\mathbf{K} \cdots \gamma_\mathbf{K})$. Now $k = k_4 \gamma_t - K \gamma_\mathbf{K}$, where $K = (\mathbf{K} \cdot \mathbf{K})^{\frac{1}{2}}$, and we have shown above that k replacing the γ_μ gives zero.[23] Hence $K \gamma_\mathbf{K}$ is equivalent to $k_4 \gamma_t$ and

$$(\gamma_t \cdots \gamma_t) - (\gamma_\mathbf{K} \cdots \gamma_\mathbf{K}) = ((K^2 - k_4^2)/K^2)(\gamma_t \cdots \gamma_t),$$

so that on multiplying by $k^{-2} d^4 k = d^4 k (k_4^2 - K^2)^{-1}$ the net effect is $-(\gamma_t \cdots \gamma_t) d^4 k / K^2$. The γ_t means just scalar waves, that is, potentials produced by charge density. The fact that $1/K^2$ does not contain k_4 means that k_4 can be integrated first, resulting in an instantaneous interaction, and the $d^3 \mathbf{K} / K^2$ is just the momentum representation of the Coulomb potential, $1/r$.

9. KLEIN GORDON EQUATION

The methods may be readily extended to particles of spin zero satisfying the Klein Gordon equation,[24]

$$\Box^2 \psi - m^2 \psi = i \partial (A_\mu \psi) / \partial x_\mu + i A_\mu \partial \psi / \partial x_\mu - A_\mu A_\mu \psi. \quad (35)$$

---

[23] A little more care is required when both γ_μ's act on the same particle. Define $x = k_4 \gamma_t + K \gamma_\mathbf{K}$, and consider $(k \cdots x) + (x \cdots k)$. Exactly this term would arise if a system, acted on by potential k carrying momentum $-k$, is disturbed by an added potential k of momentum $+k$ (the reversed sign of the momenta in the intermediate factors in the second term $x \cdots k$ has no effect since we will later integrate over all k). Hence as shown above the result is zero, but since $(k \cdots x) + (x \cdots k) = k_4^2 (\gamma_t \cdots \gamma_t) - K^2 (\gamma_\mathbf{K} \cdots \gamma_\mathbf{K})$ we can still conclude $(\gamma_\mathbf{K} \cdots \gamma_\mathbf{K}) = k_4^2 K^{-2} (\gamma_t \cdots \gamma_t)$.

[24] The equations discussed in this section were deduced from the formulation of the Klein Gordon equation given in reference 5, Section 14. The function ψ in this section has only one component and is not a spinor. An alternative formal method of making the equations valid for spin zero and also for spin 1 is (presumably) by use of the Kemmer-Duffin matrices β_μ, satisfying the commutation relation

$$\beta_\mu \beta_\nu \beta_\sigma + \beta_\sigma \beta_\nu \beta_\mu = \delta_{\mu\nu} \beta_\sigma + \delta_{\sigma\nu} \beta_\mu.$$

If we interpret a to mean $a_\mu \beta_\mu$, rather than $a_\mu \gamma_\mu$, for any a_μ, all of the equations in momentum space will remain formally identical to those for the spin $1/2$; with the exception of those in which a denominator $(p - m)^{-1}$ has been rationalized to $(p + m)(p^2 - m^2)^{-1}$ since p^2 is no longer equal to a number, $p \cdot p$. But p^2 does equal $(p \cdot p)p$ so that $(p - m)^{-1}$ may now be interpreted as $(mp + m^2 + p^2 - p \cdot p)(p \cdot p - m^2)^{-1} m^{-1}$. This implies that equations in coordinate space will be valid of the function $K_+(2, 1)$ is given as $K_+(2, 1) = [(i\nabla_2 + m) - m^{-1}(\nabla_2^2 + \Box_2^2)] I_+(2, 1)$ with $\nabla_2 = \beta_\mu \partial / \partial x_{2\mu}$. This is all in virtue of the fact that the many component wave function ψ (5 components for spin 0, 10 for spin 1) satisfies $(i\nabla - m)\psi = A\psi$ which is formally identical to the Dirac Equation. See W. Pauli, Rev. Mod. Phys. 13, 203 (1940).

The important kernel is now $I_+(2, 1)$ defined in (I, Eq. (32)). For a free particle, the wave function $\psi(2)$ satisfies $+\Box^2\psi - m^2\psi = 0$. At a point, 2, inside a space time region it is given by

$$\psi(2) = \int [\psi(1)\partial I_+(2, 1)/\partial x_{1\mu} - (\partial\psi/\partial x_{1\mu})I_+(2, 1)]N_\mu(1)d^3V_1,$$

(as is readily shown by the usual method of demonstrating Green's theorem) the integral being over an entire 3-surface boundary of the region (with normal vector N_μ). Only the positive frequency components of ψ contribute from the surface preceding the time corresponding to 2, and only negative frequencies from the surface future to 2. These can be interpreted as electrons and positrons in direct analogy to the Dirac case.

The right-hand side of (35) can be considered as a source of new waves and a series of terms written down to represent matrix elements for processes of increasing order. There is only one new point here, the term in $A_\mu A_\mu$ by which two quanta can act at the same time. As an example, suppose three quanta or potentials, $a_\mu \exp(-iq_a \cdot x)$, $b_\mu \exp(-iq_b \cdot x)$, and $c_\mu \exp(-iq_c \cdot x)$ are to act in that order on a particle of original momentum $p_{0\mu}$ so that $p_a = p_0 + q_a$ and $p_b = p_a + q_b$; the final momentum being $p_c = p_b + q_c$. The matrix element is the sum of three terms ($p^2 = p_\mu p_\mu$) (illustrated in Fig. 7).

$$(p_c \cdot c + p_b \cdot c)(p_b^2 - m^2)^{-1}(p_b \cdot b + p_a \cdot b)$$
$$\times (p_a^2 - m^2)^{-1}(p_a \cdot a + p_0 \cdot a)$$
$$-(p_c \cdot c + p_b \cdot c)(p_b^2 - m^2)^{-1}(b \cdot a) \quad (36)$$
$$-(c \cdot b)(p_a^2 - m^2)^{-1}(p_a \cdot a + p_0 \cdot a).$$

The first comes when each potential acts through the perturbation $i\partial(A_\mu\psi)/\partial x_\mu + iA_\mu\partial\psi/\partial x_\mu$. These gradient operators in momentum space mean respectively the momentum after and before the potential A_μ operates. The second term comes from b_μ and a_μ acting at the same instant and arises from the $A_\mu A_\mu$ term in (a). Together b_μ and a_μ carry momentum $q_{b\mu} + q_{a\mu}$ so that after $b \cdot a$ operates the momentum is $p_0 + q_a + q_b$ or p_b. The final term comes from c_μ and b_μ operating together in a similar manner. The term $A_\mu A_\mu$ thus permits a new type of process in which two quanta can be emitted (or absorbed, or one absorbed, one emitted) at the same time. There is no $a \cdot c$ term for the order a, b, c we have assumed. In an actual problem there would be other terms like (36) but with alterations in the order in which the quanta a, b, c act. In these terms $a \cdot c$ would appear.

As a further example the self-energy of a particle of momentum p_μ is

$$(e^2/2\pi im)\int [(2p-k)_\mu((p-k)^2 - m^2)^{-1}$$
$$\times (2p-k)_\mu - \delta_{\mu\mu}]d^4k k^{-2}C(k^2),$$

where the $\delta_{\mu\mu} = 4$ comes from the $A_\mu A_\mu$ term and represents the possibility of the simultaneous emission and absorption of the same virtual quantum. This integral without the $C(k^2)$ diverges quadratically and would not converge if $C(k^2) = -\lambda^2/(k^2 - \lambda^2)$. Since the interaction occurs through the gradients of the potential, we must use a stronger convergence factor, for example $C(k^2) = \lambda^4(k^2 - \lambda^2)^{-2}$, or in general (17) with $\int_0^\infty \lambda^2 G(\lambda) d\lambda = 0$. In this case the self-energy converges but depends quadratically on the cut-off λ and is not necessarily small compared to m. The radiative corrections to scattering after mass renormalization are insensitive to the cut-off just as for the Dirac equation.

When there are several particles one can obtain Bose statistics by the rule that if two processes lead to the same state but with two electrons exchanged, their amplitudes are to be added (rather than subtracted as for Fermi statistics). In this case equivalence to the second quantization treatment of Pauli and Weisskopf should be demonstrable in a way very much like that given in I (appendix) for Dirac electrons. The Bose statistics mean that the sign of contribution of a closed loop to the vacuum polarization is the opposite of what it is for the Fermi case (see I). It is ($p_b = p_a + q$)

$$J_{\mu\nu} = \frac{e^2}{2\pi im}\int [(p_{b\mu} + p_{a\mu})(p_{b\nu} + p_{a\nu})(p_a^2 - m^2)^{-1}$$
$$\times (p_b^2 - m^2)^{-1} - \delta_{\mu\nu}(p_a^2 - m^2)^{-1}$$
$$- \delta_{\mu\nu}(p_b^2 - m^2)^{-1}]d^4p_a$$

giving,

$$J_{\mu\nu}{}^P = \frac{e^2}{\pi}(q_\mu q_\nu - \delta_{\mu\nu}q^2)\left[\frac{1}{6}\ln\frac{\lambda^2}{m^2} + \frac{1}{9} + \frac{4m^2 - q^2}{3q^2}\left(1 - \frac{\theta}{\tan\theta}\right)\right],$$

the notation as in (33). The imaginary part for $(q^2)^{\frac{1}{2}} > 2m$ is again positive representing the loss in the probability of finding the final state to be a vacuum, associated with the possibilities of pair production. Fermi statistics would give a gain in probability (and also a charge renormalization of opposite sign to that expected).

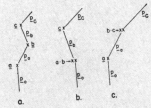

FIG. 7. Klein-Gordon particle in three potentials, Eq. (36). The coupling to the electromagnetic field is now, for example, $p_b \cdot a + p_a \cdot a$, and a new possibility arises, (b), of simultaneous interaction with two quanta $a \cdot b$. The propagation factor is now $(p \cdot p - m^2)^{-1}$ for a particle of momentum p_μ.

10. APPLICATION TO MESON THEORIES

The theories which have been developed to describe mesons and the interaction of nucleons can be easily expressed in the language used here. Calculations, to lowest order in the interactions can be made very easily for the various theories, but agreement with experimental results is not obtained. Most likely all of our present formulations are quantitatively unsatisfactory. We shall content ourselves therefore with a brief summary of the methods which can be used.

The nucleons are usually assumed to satisfy Dirac's equation so that the factor for propagation of a nucleon of momentum p is $(p-M)^{-1}$ where M is the mass of the nucleon (which implies that nucleons can be created in pairs). The nucleon is then assumed to interact with mesons, the various theories differing in the form assumed for this interaction.

First, we consider the case of neutral mesons. The theory closest to electrodynamics is the theory of vector mesons with vector coupling. Here the factor for emission or absorption of a meson is $g\gamma_\mu$ when this meson is "polarized" in the μ direction. The factor g, the "mesonic charge," replaces the electric charge e. The amplitude for propagation of a meson of momentum q in intermediate states is $(q^2-\mu^2)^{-1}$ (rather than q^{-2} as it is for light) where μ is the mass of the meson. The necessary integrals are made finite by convergence factors $C(q^2-\mu^2)$ as in electrodynamics. For scalar mesons with scalar coupling the only change is that one replaces the γ_μ by 1 in emission and absorption. There is no longer a direction of polarization, μ, to sum upon. For pseudoscalar mesons, pseudoscalar coupling replace γ_μ by $\gamma_5 = i\gamma_x\gamma_y\gamma_z\gamma_t$. For example, the self-energy matrix of a nucleon of momentum p in this theory is

$$(g^2/\pi i)\int \gamma_5(p-k-M)^{-1}\gamma_5 d^4k(k^2-\mu^2)^{-1}C(k^2-\mu^2).$$

Other types of meson theory result from the replacement of γ_μ by other expressions (for example by $\frac{1}{2}(\gamma_\mu\gamma_\nu - \gamma_\nu\gamma_\mu)$ with a subsequent sum over all μ and ν for virtual mesons). Scalar mesons with vector coupling result from the replacement of γ_μ by $\mu^{-1}q$ where q is the final momentum of the nucleon minus its initial momentum, that is, it is the momentum of the meson if absorbed, or the negative of the momentum of a meson emitted. As is well known, this theory with neutral mesons gives zero for all processes, as is proved by our discussion on longitudinal waves in electrodynamics. Pseudoscalar mesons with pseudo-vector coupling corresponds to γ_μ being replaced by $\mu^{-1}\gamma_5 q$ while vector mesons with tensor coupling correspond to using $(2\mu)^{-1}(\gamma_\mu q - q\gamma_\mu)$. These extra gradients involve the danger of producing higher divergencies for real processes. For example, $\gamma_5 q$ gives a logarithmically divergent interaction of neutron and electron.[25] Although these divergencies can be held by strong enough convergence factors, the results then are sensitive to the method used for convergence and the size of the cut-off values of λ. For low order processes $\mu^{-1}\gamma_5 q$ is equivalent to the pseudoscalar interaction $2M\mu^{-1}\gamma_5$ because if taken between free particle wave functions of the nucleon of momenta p_1 and $p_2 = p_1 + q$, we have

$$(\bar{u}_2\gamma_5 q u_1) = (\bar{u}_2\gamma_5(p_2-p_1)u_1) = -(\bar{u}_2 p_2\gamma_5 u_1) \\ -(\bar{u}_2\gamma_5 p_1 u_1) = -2M(\bar{u}_2\gamma_5 u_1)$$

since γ_5 anticommutes with p_2 and p_2 operating on the state 2 equivalent to M as is p_1 on the state 1. This shows that the γ_5 interaction is unusually weak in the non-relativistic limit (for example the expected value of γ_5 for a free nucleon is zero), but since $\gamma_5^2=1$ is not small, pseudoscalar theory gives a more important interaction in second order than it does in first. Thus the pseudoscalar coupling constant should be chosen to fit nuclear forces including these important second order processes.[26] The equivalence of pseudoscalar and pseudovector coupling which holds for low order processes therefore does not hold when the pseudoscalar theory is giving its most important effects. These theories will therefore give quite different results in the majority of practical problems.

In calculating the corrections to scattering of a nucleon by a neutral vector meson field (γ_μ) due to the effects of virtual mesons, the situation is just as in electrodynamics, in that the result converges without need for a cut-off and depends only on gradients of the meson potential. With scalar (1) or pseudoscalar (γ_5) neutral mesons the result diverges logarithmically and so must be cut off. The part sensitive to the cut-off, however, is directly proportional to the meson potential. It may thereby be removed by a renormalization of mesonic charge g. After this renormalization the results depend only on gradients of the meson potential and are essentially independent of cut-off. This is in addition to the mesonic charge renormalization coming from the production of virtual nucleon pairs by a meson, analogous to the vacuum polarization in electrodynamics. But here there is a further difference from electrodynamics for scalar or pseudoscalar mesons in that the polarization also gives a term in the induced current proportional to the meson potential representing therefore an additional renormalization of the *mass of the meson* which usually depends quadratically on the cut-off.

Next consider charged mesons in the absence of an electromagnetic field. One can introduce isotopic spin operators in an obvious way. (Specifically replace the neutral γ_5, say, by $\tau_i\gamma_5$ and sum over $i=1, 2$ where $\tau_1 = \tau_+ + \tau_-$, $\tau_2 = i(\tau_+ - \tau_-)$ and τ_+ changes neutron to proton (τ_+ on proton=0) and τ_- changes proton to neutron.) It is just as easy for practical problems simply to keep track of whether the particle is a proton or a neutron on a diagram drawn to help write down the

---

[25] M. Slotnick and W. Heitler, Phys. Rev. 75, 1645 (1949).

[26] H. A. Bethe, Bull. Am. Phys. Soc. 24, 3, Z3 (Washington, 1949).

matrix element. This excludes certain processes. For example in the scattering of a negative meson from q_1 to q_2 by a neutron, the meson q_2 must be emitted first (in order of operators, not time) for the neutron cannot absorb the negative meson q_1 until it becomes a proton. That is, in comparison to the Klein Nishina formula (15), only the analogue of second term (see Fig. 5(b)) would appear in the scattering of negative mesons by neutrons, and only the first term (Fig. 5(a)) in the neutron scattering of positive mesons.

The source of mesons of a given charge is not conserved, for a neutron capable of emitting negative mesons may (on emitting one, say) become a proton no longer able to do so. The proof that a perturbation q gives zero, discussed for longitudinal electromagnetic waves, fails. This has the consequence that vector mesons, if represented by the interaction γ_μ would not satisfy the condition that the divergence of the potential is zero. The interaction is to be taken[27] as $\gamma_\mu - \mu^{-2} q_\mu q$ in emission and as γ_μ in absorption if the real emission of mesons with a non-zero divergence of potential is to be avoided. (The correction term $\mu^{-2} q_\mu q$ gives zero in the neutral case.) The asymmetry in emission and absorption is only apparent, as this is clearly the same thing as subtracting from the original $\gamma_\mu \cdots \gamma_\mu$, a term $\mu^{-2} q \cdots q$. That is, if the term $-\mu^{-2} q_\mu q$ is omitted the resulting theory describes a combination of mesons of spin one and spin zero. The spin zero mesons, coupled by vector coupling q, are removed by subtracting the term $\mu^{-2} q \cdots q$.

The two extra gradients $q \cdots q$ make the problem of diverging integrals still more serious (for example the interaction between two protons corresponding to the exchange of two charged vector mesons depends quadratically on the cut-off if calculated in a straightforward way). One is tempted in this formulation to choose simply $\gamma_\mu \cdots \gamma_\mu$ and accept the admixture of spin zero mesons. But it appears that this leads in the conventional formalism to negative energies for the spin zero component. This shows one of the advantages of the

---

[27] The vector meson field potentials φ_μ satisfy
$$-\partial/\partial x_\nu (\partial \varphi_\mu / \partial x_\nu - \partial \varphi_\nu / \partial x_\mu) - \mu^2 \varphi_\mu = -4\pi s_\mu,$$
where s_μ, the source for such mesons, is the matrix element of γ_μ between states of neutron and proton. By taking the divergence $\partial/\partial x_\mu$ of both sides, conclude that $\partial \varphi_\nu / \partial x_\nu = 4\pi \mu^{-2} \partial s_\nu / \partial x_\nu$ so that the original equation can be rewritten as
$$\Box^2 \varphi_\mu - \mu^2 \varphi_\mu = -4\pi (s_\mu + \mu^{-2} \partial/\partial x_\mu (\partial s_\nu / \partial x_\nu)).$$
The right hand side gives in momentum representation $\gamma_\mu - \mu^{-2} q_\mu q_\nu \gamma_\nu$, the left yields the $(q^2 - \mu^2)^{-1}$ and finally the interaction $s_\mu \varphi_\mu$ in the Lagrangian gives the γ_μ on absorption.
Proceeding in this way find generally that particles of spin one can be represented by a four-vector u_μ (which, for a free particle of momentum q satisfies $q \cdot u = 0$). The propagation of virtual particles of momentum q from state ν to μ is represented by multiplication by the 4–4 matrix (or tensor) $P_{\mu\nu} = (\delta_{\mu\nu} - \mu^{-2} q_\mu q_\nu) \times (q^2 - \mu^2)^{-1}$. The first-order interaction (from the Proca equation) with an electromagnetic potential $a \exp(-ik \cdot x)$ corresponds to multiplication by the matrix $E_{\mu\nu} = (q_2 \cdot a + q_1 \cdot a) \delta_{\mu\nu} - q_{2\mu} a_\nu - q_{1\nu} a_\mu$, where q_1 and $q_2 = q_1 + k$ are the momenta before and after the interaction. Finally, two potentials a, b may act simultaneously, with matrix $E'_{\mu\nu} = -(a \cdot b) \delta_{\mu\nu} + b_\mu a_\nu$.

---

method of second quantization of meson fields over the present formulation. There such errors of sign are obvious while here we seem to be able to write seemingly innocent expressions which can give absurd results. Pseudovector mesons with pseudovector coupling correspond to using $\gamma_5(\gamma_\mu - \mu^{-2} q_\mu q)$ for absorption and $\gamma_5 \gamma_\mu$ for emission for both charged and neutral mesons.

In the presence of an electromagnetic field, whenever the nucleon is a proton it interacts with the field in the way described for electrons. The meson interacts in the scalar or pseudoscalar case as a particle obeying the Klein-Gordon equation. It is important here to use the method of calculation of Bethe and Pauli, that is, a virtual meson is assumed to have the same "mass" during all its interactions with the electromagnetic field. The result for mass μ and for $(\mu^2 + \lambda^2)^{\frac{1}{2}}$ are subtracted and the difference integrated over the function $G(\lambda) d\lambda$. A separate convergence factor is not provided for each meson propagation between electromagnetic interactions, otherwise gauge invariance is not insured. When the coupling involves a gradient, such as $\gamma_5 q$ where q is the final minus the initial momentum of the nucleon, the vector potential A must be subtracted from the momentum of the proton. That is, there is an additional coupling $\pm \gamma_5 A$ (plus when going from proton to neutron, minus for the reverse) representing the new possibility of a simultaneous emission (or absorption) of meson and photon.

Emission of positive or absorption of negative virtual mesons are represented in the same term, the sign of the charge being determined by temporal relations as for electrons and positrons.

Calculations are very easily carried out in this way to lowest order in g^2 for the various theories for nucleon interaction, scattering of mesons by nucleons, meson production by nuclear collisions and by gamma-rays, nuclear magnetic moments, neutron electron scattering, etc., However, no good agreement with experiment results, when these are available, is obtained. Probably all of the formulations are incorrect. An uncertainty arises since the calculations are only to first order in g^2, and are not valid if $g^2/\hbar c$ is large.

The author is particularly indebted to Professor H. A. Bethe for his explanation of a method of obtaining finite and gauge invariant results for the problem of vacuum polarization. He is also grateful for Professor Bethe's criticisms of the manuscript, and for innumerable discussions during the development of this work. He wishes to thank Professor J. Ashkin for his careful reading of the manuscript.

APPENDIX

In this appendix a method will be illustrated by which the simpler integrals appearing in problems in electrodynamics can be directly evaluated. The integrals arising in more complex processes lead to rather complicated functions, but the study of the relations of one integral to another and their expression in terms of simpler integrals may be facilitated by the methods given here.

As a typical problem consider the integral (12) appearing in the first order radiationless scattering problem:

$$\int \gamma_\mu(p_2-k-m)^{-1}a(p_1-k-m)^{-1}\gamma_\mu k^{-2}d^4kC(k^2), \quad (1a)$$

where we shall take $C(k^2)$ to be typically $-\lambda^2(k^2-\lambda^2)^{-1}$ and d^4k means $(2\pi)^{-2}dk_1dk_2dk_3dk_4$. We first rationalize the factors $(p-k-m)^{-1}=(p-k+m)((p-k)^2-m^2)^{-1}$ obtaining,

$$\int \gamma_\mu(p_2-k+m)a(p_1-k+m)\gamma_\mu k^{-2}d^4kC(k^2)$$
$$\times ((p_1-k)^2-m^2)^{-1}((p_2-k)^2-m^2)^{-1}. \quad (2a)$$

The matrix expression may be simplified. It appears to be best to do so *after* the integrations are performed. Since $AB=2A\cdot B-BA$ where $A\cdot B=A_\mu B_\mu$ is a number commuting with all matrices, find, if R is any expression, and A a vector, since $\gamma_\mu A = -A\gamma_\mu+2A_\mu$,

$$\gamma_\mu A R \gamma_\mu = -A\gamma_\mu R\gamma_\mu+2RA. \quad (3a)$$

Expressions between two γ_μ's can be thereby reduced by induction. Particularly useful are

$$\gamma_\mu\gamma_\mu=4$$
$$\gamma_\mu A\gamma_\mu=-2A$$
$$\gamma_\mu AB\gamma_\mu=2(AB+BA)=4A\cdot B \quad (4a)$$
$$\gamma_\mu ABC\gamma_\mu=-2CBA$$

where A, B, C are any three vector-matrices (i.e., linear combinations of the four γ's).

In order to calculate the integral in (2a) the integral may be written as the sum of three terms (since $k=k_\sigma\gamma_\sigma$),

$$\gamma_\mu(p_2+m)a(p_1+m)\gamma_\mu J_1-[\gamma_\mu\gamma_\sigma a(p_1+m)\gamma_\mu$$
$$+\gamma_\mu(p_2+m)a\gamma_\sigma\gamma_\mu]J_2+\gamma_\mu\gamma_\sigma a\gamma_\tau\gamma_\mu J_3, \quad (5a)$$

where

$$J_{(1;2;3)}=\int (1; k_\sigma; k_\sigma k_\tau)k^{-2}d^4kC(k^2)$$
$$\times ((p_2-k)^2-m^2)^{-1}((p_1-k)^2-m^2)^{-1}. \quad (6a)$$

That is for J_1 the $(1; k_\sigma; k_\sigma k_\tau)$ is replaced by 1, for J_2 by k_σ, and for J_3 by $k_\sigma k_\tau$.

More complex processes of the first order involve more factors like $((p_3-k)^2-m^2)^{-1}$ and a corresponding increase in the number of k's which may appear in the numerator, as $k_\sigma k_\tau k_\nu \cdots$. Higher order processes involving two or more virtual quanta involve similar integrals but with factors possibly involving $k+k'$ instead of just k, and the integral extending on $k^{-2}d^4kC(k^2)k'^{-2}d^4k'C(k'^2)$. They can be simplified by methods analogous to those used on the first order integrals.

The factors $(p-k)^2-m^2$ may be written

$$(p-k)^2-m^2=k^2-2p\cdot k-\Delta, \quad (7a)$$

where $\Delta=m^2-p^2$, $\Delta_1=m_1^2-p_1^2$, etc., and we can consider dealing with cases of greater generality in that the different denominators need not have the same value of the mass m. In our specific problem (6a), $p_1^2=m^2$ so that $\Delta_1=0$, but we desire to work with greater generality.

Now for the factor $C(k^2)/k^2$ we shall use $-\lambda^2(k^2-\lambda^2)^{-1}k^{-2}$. This can be written as

$$-\lambda^2/(k^2-\lambda^2)k^2=k^{-2}C(k^2)=-\int_0^{\lambda^2}dL(k^2-L)^{-2}. \quad (8a)$$

Thus we can replace $k^{-2}C(k^2)$ by $(k^2-L)^{-2}$ and at the end integrate the result with respect to L from zero to λ^2. We can for many practical purposes consider λ^2 very large relative to m^2 or p^2. When the original integral converges even without the convergence factor, it will be obvious since the L integration will then be convergent to infinity. If an infra-red catastrophe exists in the integral one can simply assume quanta have a small mass λ_{min} and extend the integral on L from λ^2_{min} to λ^2, rather than from zero to λ^2.

We then have to do integrals of the form

$$\int (1; k_\sigma; k_\sigma k_\tau)d^4k(k^2-L)^{-2}(k^2-2p_1\cdot k-\Delta_1)^{-1}$$
$$\times (k^2-2p_2\cdot k-\Delta_2)^{-1}, \quad (9a)$$

where by $(1; k_\sigma; k_\sigma k_\tau)$ we mean that in the place of this symbol either 1, or k_σ, or $k_\sigma k_\tau$ may stand in different cases. In more complicated problems there may be more factors $(k^2-2p_3\cdot k-\Delta_3)^{-1}$ or other powers of these factors (the $(k^2-L)^{-2}$ may be considered as a special case of such a factor with $p_i=0$, $\Delta_i=L$) and further factors like $k_\sigma k_\tau k_\rho \cdots$ in the numerator. The poles in all the factors are made definite by the assumption that L, and the Δ's have infinitesimal negative imaginary parts.

We shall do the integrals of successive complexity by induction. We start with the simplest convergent one, and show

$$\int d^4k(k^2-L)^{-3}=(8iL)^{-1}. \quad (10a)$$

For this integral is $\int (2\pi)^{-2}dk_4 d^3K(k_4^2-\mathbf{K}\cdot\mathbf{K}-L)^{-3}$ where the vector $\mathbf{K}$, of magnitude $K=(\mathbf{K}\cdot\mathbf{K})^{\frac{1}{2}}$ is k_1, k_2, k_3. The integral on k_4 shows third order poles at $k_4=+(K^2+L)^{\frac{1}{2}}$ and $k_4=-(K^2+L)^{\frac{1}{2}}$. Imagining, in accordance with our definitions, that L has a small negative imaginary part only the first is below the real axis. The contour can be closed by an infinite semi-circle below this axis, without change of the value of the integral since the contribution from the semi-circle vanishes in the limit. Thus the contour can be shrunk about the pole $k_4=+(K^2+L)^{\frac{1}{2}}$ and the resulting k_4 integral is $-2\pi i$ times the residue at this pole. Writing $k_4=(K^2+L)^{\frac{1}{2}}+\epsilon$ and expanding $(k_4^2-K^2-L)^{-3}=\epsilon^{-3}(\epsilon+2(K^2+L)^{\frac{1}{2}})^{-3}$ in powers of ϵ, the residue, being the coefficient of the term ϵ^{-1}, is seen to be $6(2(K^2+L)^{\frac{1}{2}})^{-5}$ so our integral is

$$-(3i/32\pi)\int_0^\infty 4\pi K^2 dK(K^2+L)^{-5/2}=(3/8i)(1/3L)$$

establishing (10a).

We also have $\int k_\sigma d^4k(k^2-L)^{-3}=0$ from the symmetry in the k space. We write these results as

$$(8i)\int (1; k_\sigma)d^4k(k^2-L)^{-3}=(1; 0)L^{-1}, \quad (11a)$$

where in the brackets $(1; k_\sigma)$ and $(1; 0)$ corresponding entries are to be used.

Substituting $k=k'-p$ in (11a), and calling $L-p^2=\Delta$ shows that

$$(8i)\int (1; k_\sigma)d^4k(k^2-2p\cdot k-\Delta)^{-3}=(1; p_\sigma)(p^2+\Delta)^{-1}. \quad (12a)$$

By differentiating both sides of (12a) with respect to Δ, or with respect to p_τ there follows directly

$$(24i)\int (1; k_\sigma; k_\sigma k_\tau)d^4k(k^2-2p\cdot k-\Delta)^{-4}$$
$$=-(1; p_\sigma; p_\sigma p_\tau-\frac{1}{2}\delta_{\sigma\tau}(p^2+\Delta))(p^2+\Delta)^{-2}. \quad (13a)$$

Further differentiations give directly successive integrals including more k factors in the numerator and higher powers of $(k^2-2p\cdot k-\Delta)$ in the denominator.

The integrals so far only contain one factor in the denominator. To obtain results for two factors we make use of the identity

$$a^{-1}b^{-1}=\int_0^1 dx(ax+b(1-x))^{-2}, \quad (14a)$$

(suggested by some work of Schwinger's involving Gaussian integrals). This represents the product of two reciprocals as a parametric integral over one and will therefore permit integrals with two factors to be expressed in terms of one. For other powers of a, b, we make use of all of the identities, such as

$$a^{-2}b^{-1}=\int_0^1 2xdx(ax+b(1-x))^{-3}, \quad (15a)$$

deducible from (14a) by successive differentiations with respect to a or b.

To perform an integral, such as

$$(8i)\int (1; k_\sigma)d^4k(k^2-2p_1\cdot k-\Delta_1)^{-2}(k^2-2p_2\cdot k-\Delta_2)^{-1}, \quad (16a)$$

write, using (15a),

$$(k^2-2p_1\cdot k-\Delta_1)^{-2}(k^2-2p_2\cdot k-\Delta_2)^{-1} = \int_0^1 2xdx(k^2-2p_x\cdot k-\Delta_x)^{-3},$$

where

$$p_x = xp_1 + (1-x)p_2 \quad \text{and} \quad \Delta_x = x\Delta_1 + (1-x)\Delta_2, \quad (17a)$$

(note that Δ_x is *not* equal to $m^2 - p_x^2$ so that the expression (16a) is $(8i)\int_0^1 2xdx\int(1;k_\sigma)d^4k(k^2-2p_x\cdot k-\Delta_x)^{-3}$ which may now be evaluated by (12a) and is

$$(16a) = \int_0^1 (1; p_{x\sigma})2xdx(p_x^2+\Delta_x)^{-1}, \quad (18a)$$

where p_x, Δ_x are given in (17a). The integral in (18a) is elementary, being the integral of ratio of polynomials, the denominator of second degree in x. The general expression although readily obtained is a rather complicated combination of roots and logarithms.

Other integrals can be obtained again by parametric differentiation. For example differentiation of (16a), (18a) with respect to Δ_2 or $p_{2\tau}$ gives

$$(8i)\int(1; k_\sigma; k_\sigma k_\tau)d^4k(k^2-2p_1\cdot k-\Delta_1)^{-2}(k^2-2p_2\cdot k-\Delta_2)^{-2}$$

$$= -\int_0^1 (1; p_{x\sigma}; p_{x\sigma}p_{x\tau} - \tfrac{1}{2}\delta_{\sigma\tau}(p_x^2+\Delta_x))$$
$$\times 2x(1-x)dx(p_x^2+\Delta_x)^{-2}, \quad (19a)$$

again leading to elementary integrals.

As an example, consider the case that the second factor is just $(k^2-L)^{-2}$ and in the first put $p_1=p$, $\Delta_1=\Delta$. Then $p_x = xp$, $\Delta_x = x\Delta + (1-x)L$. There results

$$(8i)\int(1; k_\sigma; k_\sigma k_\tau)d^4k(k^2-L)^{-2}(k^2-2p\cdot k-\Delta)^{-2}$$

$$= -\int_0^1 (1; xp_\sigma; x^2p_\sigma p_\tau - \tfrac{1}{2}\delta_{\sigma\tau}(x^2p^2+\Delta_x))$$
$$\times 2x(1-x)dx(x^2p^2+\Delta_x)^{-2}. \quad (20a)$$

Integrals with three factors can be reduced to those involving two by using (14a) again. They, therefore, lead to integrals with two parameters (e.g., see application to radiative correction to scattering below).

The methods of calculation given in this paper are deceptively simple when applied to the lower order processes. For processes of increasingly higher orders the complexity and difficulty increases rapidly, and these methods soon become impractical in their present form.

A. Self-Energy

The self-energy integral (19) is

$$(e^2/\pi i)\int \gamma_\mu(p-k-m)^{-1}\gamma_\mu k^{-2}d^4k C(k^2), \quad (19)$$

so that it requires that we find (using the principle of (8a)) the integral on L from 0 to λ^2 of

$$\int \gamma_\mu(p-k+m)\gamma_\mu d^4k(k^2-L)^{-2}(k^2-2p\cdot k)^{-1},$$

since $(p-k)^2-m^2 = k^2-2p\cdot k$, as $p^2=m^2$. This is of the form (16a) with $\Delta_1 = L$, $p_1 = 0$, $\Delta_2 = 0$, $p_2 = p$ so that (18a) gives, since $p_x = (1-x)p$, $\Delta_x = xL$,

$$(8i)\int(1; k_\sigma)d^4k(k^2-L)^{-2}(k^2-2p\cdot k)^{-1}$$
$$= \int_0^1 (1; (1-x)p_\sigma)2xdx((1-x)^2m^2+xL)^{-1},$$

or performing the integral on L, as in (8),

$$(8i)\int(1; k_\sigma)d^4k k^{-2}C(k^2)(k^2-2p\cdot k)^{-1}$$
$$= \int_0^1 (1; (1-x)p_\sigma)2dx \ln\frac{x\lambda^2+(1-x)^2m^2}{(1-x)^2m^2}.$$

Assuming now that $\lambda^2\gg m^2$ we neglect $(1-x)^2m^2$ relative to $x\lambda^2$ in the argument of the logarithm, which then becomes $(\lambda^2/m^2)(x/(1-x)^2)$. Then since $\int_0^1 dx \ln(x(1-x)^{-2}) = 1$ and

$\int_0^1(1-x)dx \ln(x(1-x)^{-2}) = -(1/4)$ find

$$(8i)\int(1; k_\sigma)k^{-2}C(k^2)d^4k(k^2-2p\cdot k)^{-1}$$
$$= \left(2\ln\frac{\lambda^2}{m^2}+2; p_\sigma\left(\ln\frac{\lambda^2}{m^2}-\tfrac{1}{2}\right)\right),$$

so that substitution into (19) (after the $(p-k-m)^{-1}$ in (19) is replaced by $(p-k+m)(k^2-2p\cdot k)^{-1}$) gives

$$(19) = (e^2/8\pi)\gamma_\mu[(p+m)(2\ln(\lambda^2/m^2)+2)$$
$$-p(\ln(\lambda^2/m^2)-\tfrac{1}{2})]\gamma_\mu$$
$$= (e^2/8\pi)[8m(\ln(\lambda^2/m^2)+1)-p(2\ln(\lambda^2/m^2)+5)], \quad (20)$$

using (4a) to remove the γ_μ's. This agrees with Eq. (20) of the text, and gives the self-energy (21) when p is replaced by m.

B. Corrections to Scattering

The term (12) in the radiationless scattering, after rationalizing the matrix denominators and using $p_1^2 = p_2^2 = m^2$ requires the integrals (9a), as we have discussed. This is an integral with three denominators which we do in two stages. First the factors $(k^2-2p_1\cdot k)$ and $(k^2-2p_2\cdot k)$ are combined by a parameter y;

$$(k^2-2p_1\cdot k)^{-1}(k^2-2p_2\cdot k)^{-1} = \int_0^1 dy(k^2-2p_y\cdot k)^{-2},$$

from (14a) where

$$p_y = yp_1 + (1-y)p_2. \quad (21a)$$

We therefore need the integrals

$$(8i)\int(1; k_\sigma; k_\sigma k_\tau)d^4k(k^2-L)^{-2}(k^2-2p_y\cdot k)^{-2}, \quad (22a)$$

which we will then integrate with respect to y from 0 to 1. Next we do the integrals (22a) immediately from (20a) with $p = p_y$, $\Delta = 0$:

$$(22a) = -\int_0^1 \int_0^1 (1; xp_{y\sigma}; x^2p_{y\sigma}p_{y\tau}$$
$$-\tfrac{1}{2}\delta_{\sigma\tau}(x^2p_y^2+(1-x)L))2x(1-x)dx(x^2p_y^2+L(1-x))^{-2}dy.$$

We now turn to the integrals on L as required in (8a). The first term, (1), in $(1; k_\sigma; k_\sigma k_\tau)$ gives no trouble for large L, but if L is put equal to zero there results $x^{-2}p_y^{-2}$ which leads to a diverging integral on x as $x\to 0$. This infra-red catastrophe is analyzed by using $\lambda_{\min}^2$ for the lower limit of the L integral. For the last term the upper limit of L must be kept as λ^2. Assuming $\lambda_{\min}^2 \ll p_y^2 \ll \lambda^2$ the x integrals which remain are trivial, as in the self-energy case. One finds

$$-(8i)\int(k^2-\lambda_{\min}^2)^{-1}d^4k C(k^2-\lambda_{\min}^2)(k^2-2p_1\cdot k)^{-1}(k^2-2p_2\cdot k)^{-1}$$
$$= \int_0^1 p_y^{-2}dy \ln(p_y^2/\lambda_{\min}^2) \quad (23a)$$

$$-(8i)\int k_\sigma k^{-2}d^4k C(k^2)(k^2-2p_1\cdot k)^{-1}(k^2-2p_2\cdot k)^{-1}$$
$$= 2\int_0^1 p_{y\sigma}p_y^{-2}dy, \quad (24a)$$

$$-(8i)\int k_\sigma k_\tau k^{-2}d^4k C(k^2)(k^2-2p_1\cdot k)^{-1}(k^2-2p_2\cdot k)^{-1}$$
$$= \int_0^1 p_{y\sigma}p_{y\tau}p_y^{-2}dy - \tfrac{1}{2}\delta_{\sigma\tau}\int_0^1 dy \ln(\lambda^2 p_y^{-2}) + \tfrac{1}{4}\delta_{\sigma\tau}. \quad (25a)$$

The integrals on y give,

$$\int_0^1 p_y^{-2}dy \ln(p_y^2\lambda_{\min}^{-2}) = 4(m^2\sin 2\theta)^{-1}\left[\theta \ln(m\lambda_{\min}^{-1})\right.$$
$$\left.-\int_0^\theta \alpha \tan\alpha d\alpha\right], \quad (26a)$$

$$\int_0^1 p_{y\sigma}p_y^{-2}dy = \theta(m^2\sin 2\theta)^{-1}(p_{1\sigma}+p_{2\sigma}), \quad (27a)$$

$$\int_0^1 p_{y\sigma}p_{y\tau}p_y^{-2}dy = \theta(2m^2\sin 2\theta)^{-1}(p_{1\sigma}+p_{1\tau})(p_{2\sigma}+p_{2\tau})$$
$$+q^{-2}g_{\sigma\tau}(1-\theta \operatorname{ctn}\theta), \quad (28a)$$

$$\int_0^1 dy \ln(\lambda^2 p_y^{-2}) = \ln(\lambda^2/m^2) + 2(1-\theta \operatorname{ctn}\theta). \quad (29a)$$

These integrals on y were performed as follows. Since $p_2 = p_1 + q$ where q is the momentum carried by the potential, it follows from $p_2{}^2 = p_1{}^2 = m^2$ that $2p_1 \cdot q = -q^2$ so that since $p_y = p_1 + q(1-y)$, $p_y{}^2 = m^2 - q^2 y(1-y)$. The substitution $2y - 1 = \tan\alpha/\tan\theta$ where θ is defined by $4m^2 \sin^2\theta = q^2$ is useful for it means $p_y{}^2 = m^2 \sec^2\alpha/\sec^2\theta$ and $p_y{}^{-2}dy = (m^2 \sin 2\theta)^{-1}d\alpha$ where α goes from $-\theta$ to $+\theta$.

These results are substituted into the original scattering formula (2a), giving (22). It has been simplified by frequent use of the fact that p_1 operating on the initial state is m, and likewise p_2 when it appears at the left is replaceable by m. (Thus, to simplify:

$$\gamma_\mu p_2 a p_1 \gamma_\mu = -2p_1 a p_2 \text{ by (4a)},$$
$$= -2(p_2-q)a(p_1+q) = -2(m-q)a(m+q).$$

A term like $qaq = -q^2 a + 2(a \cdot q)q$ is equivalent to just $-q^2 a$ since $q = p_2 - p_1 = m - m$ has zero matrix element.) The renormalization term requires the corresponding integrals for the special case $q = 0$.

C. Vacuum Polarization

The expressions (32) and (32') for $J_{\mu\nu}$ in the vacuum polarization problem require the calculation of the integral

$$J_{\mu\nu}(m^2) = -\frac{e^2}{\pi i}\int Sp[\gamma_\mu(p-\tfrac{1}{2}q+m)\gamma_\nu(p+\tfrac{1}{2}q+m)]d^4p$$
$$\times ((p-\tfrac{1}{2}q)^2 - m^2)^{-1}((p+\tfrac{1}{2}q)^2 - m^2)^{-1}, \quad (32)$$

where we have replaced p by $p - \tfrac{1}{2}q$ to simplify the calculation somewhat. We shall indicate the method of calculation by studying the integral,

$$I(m^2) = \int p_\sigma p_\tau d^4p ((p-\tfrac{1}{2}q)^2-m^2)^{-1}((p+\tfrac{1}{2}q)^2-m^2)^{-1}.$$

The factors in the denominator, $p^2 - p \cdot q - m^2 + \tfrac{1}{4}q^2$ and $p^2 + p \cdot q - m^2 + \tfrac{1}{4}q^2$ are combined as usual by (8a) but for symmetry we substitute $x = \tfrac{1}{2}(1+\eta)$, $(1-x) = \tfrac{1}{2}(1-\eta)$ and integrate η from -1 to $+1$:

$$I(m^2) = \int_{-1}^{+1} p_\sigma p_\tau d^4p(p^2 - \eta p \cdot q - m^2 + \tfrac{1}{4}q^2)^{-2}d\eta/2. \quad (30a)$$

But the integral on p will not be found in our list for it is badly divergent. However, as discussed in Section 7, Eq. (32') we do not wish $I(m^2)$ but rather $\int_0^\infty [I(m^2) - I(m^2 + \lambda^2)]G(\lambda)d\lambda$. We can calculate the difference $I(m^2) - I(m^2+\lambda^2)$ by first calculating the derivative $I'(m^2+L)$ of I with respect to m^2 at $m^2 + L$ and later integrating L from zero to λ^2. By differentiating (30a), with respect to m^2 find,

$$I'(m^2+L) = \int_{-1}^{+1} p_\sigma p_\tau d^4p (p^2 - \eta p \cdot q - m^2 - L + \tfrac{1}{4}q^2)^{-3}d\eta.$$

This still diverges, but we can differentiate again to get

$$I''(m^2+L) = 3 \int_{-1}^{+1} p_\sigma p_\tau d^4 p (p^2 - \eta p \cdot q - m^2 - L + \tfrac{1}{4}q^2)^{-4}d\eta \quad (31a)$$
$$= -(8i)^{-1}\int_{-1}^{+1}(\tfrac{1}{4}\eta^2 q_\sigma q_\tau D^{-2} - \tfrac{1}{2}\delta_{\sigma\tau}D^{-1})d\eta$$

(where $D = \tfrac{1}{4}(\eta^2-1)q^2 + m^2 + L$), which now converges and has been evaluated by (13a) with $p = \tfrac{1}{2}\eta q$ and $\Delta = m^2 + L - \tfrac{1}{4}q^2$. Now to get I' we may integrate I'' with respect to L as an indefinite integral and *we may choose any convenient arbitrary constant*. This is because a constant C in I' will mean a term $-C\lambda^2$ in $I(m^2) - I(m^2 + \lambda^2)$ which vanishes since we will integrate the results times $G(\lambda)d\lambda$ and $\int_0^\infty \lambda^2 G(\lambda)d\lambda = 0$. This means that the logarithm appearing on integrating L in (31a) presents no problem. We may take

$$I'(m^2+L) = (8i)^{-1}\int_{-1}^{+1} [\tfrac{1}{4}\eta q_\sigma q_\tau D^{-1} + \tfrac{1}{2}\delta_{\sigma\tau}\ln D]d\eta + C\delta_{\sigma\tau},$$

a subsequent integral on L and finally on η presents no new problems. There results

$$-(8i)\int p_\sigma p_\tau d^4p((p-\tfrac{1}{2}q)^2-m^2)^{-1}((p+\tfrac{1}{2}q)^2-m^2)^{-1}$$
$$= (q_\sigma q_\tau - \delta_{\sigma\tau}q^2)\left[\frac{1}{9} - \frac{4m^2-q^2}{3q^2}\left(1 - \frac{\theta}{\tan\theta}\right) + \tfrac{1}{6}\ln\frac{\lambda^2}{m^2}\right]$$
$$+ \delta_{\sigma\tau}[(\lambda^2+m^2)\ln(\lambda^2 m^{-2}+1) - C'\lambda^2], \quad (32a)$$

where we assume $\lambda^2 \gg m^2$ and have put some terms into the arbitrary constant C' which is independent of λ^2 (but in principle could depend on q^2) and which drops out in the integral on $G(\lambda)d\lambda$. We have set $q^2 = 4m^2 \sin^2\theta$.

In a very similar way the integral with m^2 in the numerator can be worked out. It is, of course, necessary to differentiate this m^2 also when calculating I' and I''. There results

$$-(8i)\int m^2 d^4p((p-\tfrac{1}{2}q)^2-m^2)^{-1}((p+\tfrac{1}{2}q)^2-m^2)^{-1}$$
$$= 4m^2(1-\theta\ctn\theta) - q^2/3 + 2(\lambda^2+m^2)\ln(\lambda^2 m^{-2}+1) - C''\lambda^2), \quad (33a)$$

with another unimportant constant C''. The complete problem requires the further integral,

$$-(8i)\int (1; p_\sigma)d^4p((p-\tfrac{1}{2}q)^2-m^2)^{-1}((p+\tfrac{1}{2}q)^2-m^2)^{-1}$$
$$= (1, 0)(4(1-\theta\ctn\theta) + 2\ln(\lambda^2 m^{-2})). \quad (34a)$$

The value of the integral (34a) times m^2 differs from (33a), of course, because the results on the right are not actually the integrals on the left, but rather equal their actual value minus their value for $m^2 = m^2 + \lambda^2$.

Combining these quantities, as required by (32), dropping the constants C', C'' and evaluating the spur gives (33). The spurs are evaluated in the usual way, noting that the spur of any odd number of γ matrices vanishes and $Sp(AB) = Sp(BA)$ for arbitrary A, B. The $Sp(1) = 4$ and we also have

$$\tfrac{1}{4}Sp[(p_1+m_1)(p_2-m_2)] = p_1 \cdot p_2 - m_1 m_2, \quad (35a)$$

$$\tfrac{1}{4}Sp[(p_1+m_1)(p_2-m_2)(p_3+m_3)(p_4-m_4)]$$
$$= (p_1 \cdot p_2 - m_1 m_2)(p_3 \cdot p_4 - m_3 m_4)$$
$$- (p_1 \cdot p_3 - m_1 m_3)(p_2 \cdot p_4 - m_2 m_4)$$
$$+ (p_1 \cdot p_4 - m_1 m_4)(p_2 \cdot p_3 - m_2 m_3), \quad (36a)$$

where p_i, m_i are arbitrary four-vectors and constants.

It is interesting that the terms of order $\lambda^2 \ln \lambda^2$ go out, so that the charge renormalization depends only logarithmically on λ^2. This is not true for some of the meson theories. Electrodynamics is suspiciously unique in the mildness of its divergence.

D. More Complex Problems

Matrix elements for complex problems can be set up in a manner analogous to that used for the simpler cases. We give three illustrations; higher order corrections to the Møller scatter-

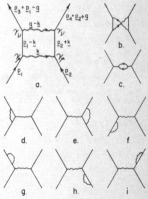

FIG. 8. The interaction between two electrons to order $(e^2/\hbar c)^2$. One adds the contribution of every figure involving two virtual quanta, Appendix D.

ing, to the Compton scattering, and the interaction of a neutron with an electromagnetic field.

For the Møller scattering, consider two electrons, one in state u_1 of momentum p_1 and the other in state u_2 of momentum p_2. Later they are found in states u_3, p_3 and u_4, p_4. This may happen (first order in $e^2/\hbar c$) because they exchange a quantum of momentum $q=p_1-p_3=p_4-p_2$ in the manner of Eq. (4) and Fig. 1. The matrix element for this process is proportional to (translating (4) to momentum space)

$$(\bar{u}_4\gamma_\mu u_2)(\bar{u}_3\gamma_\mu u_1)q^{-2}. \qquad (37a)$$

We shall discuss corrections to (37a) to the next order in $e^2/\hbar c$. (There is also the possibility that it is the electron at 2 which finally arrives at 3, the electron at 1 going to 4 through the exchange of quantum of momentum p_4-p_2. The amplitude for this process, $(\bar{u}_4\gamma_\mu u_1)(\bar{u}_3\gamma_\mu u_2)(p_3-p_2)^{-2}$, must be subtracted from (37a) in accordance with the exclusion principle. A similar situation exists to each order so that we need consider in detail only the corrections to (37a), reserving to the last the subtraction of the same terms with 3, 4 exchanged.)

One reason that (37a) is modified is that two quanta may be exchanged, in the manner of Fig. 8a. The total matrix element for all exchanges of this type is

$$(e^2/\pi i)\int(\bar{u}_3\gamma_\nu(p_1-k-m)^{-1}\gamma_\mu u_1)(\bar{u}_4\gamma_\nu(p_2+k-m)^{-1}\gamma_\mu u_2)$$
$$\cdot k^{-2}(q-k)^{-2}d^4k, \qquad (38a)$$

as is clear from the figure and the general rule that electrons of momentum p contribute in amplitude $(p-m)^{-1}$ between interactions γ_μ, and that quanta of momentum k contribute k^{-2}. In integrating on d^4k and summing over μ and ν, we add all alternatives of the type of Fig. 8a. If the time of absorption, γ_ν, of the quantum k by electron 2 is later than the absorption, γ_ν, of $q-k$, this corresponds to the virtual state p_2+k being a positron (so that (38a) contains over thirty terms of the conventional method of analysis).

In integrating over all these alternatives we have considered all possible distortions of Fig. 8a which preserve the order of events along the trajectories. We have not included the possibilities corresponding to Fig. 8b, however. Their contribution is

$$(e^2/\pi i)\int(\bar{u}_3\gamma_\nu(p_1-k-m)^{-1}\gamma_\mu u_1)$$
$$\times(\bar{u}_4\gamma_\mu(p_2+q-k-m)^{-1}\gamma_\nu u_2)k^{-2}(q-k)^{-2}d^4k, \qquad (39a)$$

as is readily verified by labeling the diagram. The contributions of all possible ways that an event can occur are to be added. This

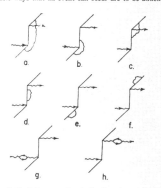

Fig. 9. Radiative correction to the Compton scattering term (a) of Fig. 5. Appendix D.

means that one adds with equal weight the integrals corresponding to each topologically distinct figure.

To this same order there are also the possibilities of Fig. 8d which give

$$(e^2/\pi i)\int(\bar{u}_3\gamma_\nu(p_3-k-m)^{-1}\gamma_\mu(p_1-k-m)^{-1}\gamma_\nu u_1)$$
$$\times(\bar{u}_4\gamma_\mu u_2)k^{-2}q^{-2}d^4k.$$

This integral on k will be seen to be precisely the integral (12) for the radiative corrections to scattering, which we have worked out. The term may be combined with the renormalization terms resulting from the difference of the effects of mass change and the terms, Figs. 8f and 8g. Figures 8e, 8h, and 8i are similarly analyzed.

Finally the term Fig. 8c is clearly related to our vacuum polarization problem, and when integrated gives a term proportional to $(\bar{u}_4\gamma_\mu u_2)(\bar{u}_3\gamma_\mu u_1)J_{\mu\nu}q^{-4}$. If the charge is renormalized the term $\ln(\lambda/m)$ in $J_{\mu\nu}$ in (33) is omitted so there is no remaining dependence on the cut-off.

The only new integrals we require are the convergent integrals (38a) and (39a). They can be simplified by rationalizing the denominators and combining them by (14a). For example (38a) involves the factors $(k^2-2p_1\cdot k)^{-1}(k^2+2p_2\cdot k)^{-1}k^{-2}(q^2+k^2-2q\cdot k)^{-2}$. The first two may be combined by (14a) with a parameter x, and the second pair by an expression obtained by differentiation (15a) with respect to b and calling the parameter y. There results a factor $(k^2-2p_x\cdot k)^{-3}(k^2+yq^2-2yq\cdot k)^{-4}$ so that the integrals on d^4k now involve two factors and can be performed by the methods given earlier in the appendix. The subsequent integrals on the parameters x and y are complicated and have not been worked out in detail.

Working with charged mesons there is often a considerable reduction of the number of terms. For example, for the interaction between protons resulting from the exchange of two mesons only the term corresponding to Fig. 8b remains. Term 8a, for example, is impossible, for if the first proton emits a positive meson the second cannot absorb it directly for only neutrons can absorb positive mesons.

As a second example, consider the radiative correction to the Compton scattering. As seen from Eq. (15) and Fig. 5 this scattering is represented by two terms, so that we can consider the corrections to each one separately. Figure 9 shows the types of terms arising from corrections to the term of Fig. 5a. Calling k the momentum of the virtual quantum, Fig. 9a gives an integral

$$\int\gamma_\mu(p_2-k-m)^{-1}e_2(p_1+q_1-k-m)^{-1}e_1(p_1-k-m)^{-1}\gamma_\mu k^{-2}d^4k,$$

convergent without cut-off and reducible by the methods outlined in this appendix.

The other terms are relatively easy to evaluate. Terms b and c of Fig. 9 are closely related to radiative corrections (although somewhat more difficult to evaluate, for one of the states is not that of a free electron, $(p_1+q)^2\neq m^2$). Terms e, f are renormalization terms. From term d must be subtracted explicitly the effect of mass Δm, as analyzed in Eqs. (26) and (27) leading to (28) with $p'=p_1+q$, $a=e_2$, $b=e_1$. Terms g, h give zero since the vacuum polarization has zero effect on free light quanta, $q_1^2=0$, $q_2^2=0$. The total is insensitive to the cut-off λ.

The result shows an infra-red catastrophe, the largest part of the effect. When cut-off at $\lambda_{\min}$, the effect proportional to $\ln(m/\lambda_{\min})$ goes as

$$(e^2/\pi)\ln(m/\lambda_{\min})(1-2\theta\operatorname{ctn}2\theta), \qquad (40a)$$

times the uncorrected amplitude, where $(p_2-p_1)^2=4m^2\sin^2\theta$. This is the same as for the radiative correction to scattering for a deflection p_2-p_1. This is physically clear since the long wave quanta are not effected by short-lived intermediate states. The infra-red effects arise[28] from a final adjustment of the field from the asymptotic coulomb field characteristic of the electron of

---

[28] F. Bloch and A. Nordsieck, Phys. Rev. **52**, 54 (1937).

momentum p_1 before the collision to that characteristic of an electron moving in a new direction p_2 after the collision.

The complete expression for the correction is a very complicated expression involving transcendental integrals.

As a final example we consider the interaction of a neutron with an electromagnetic field in virtue of the fact that the neutron may emit a virtual negative meson. We choose the example of pseudoscalar mesons with pseudovector coupling. The change in amplitude due to an electromagnetic field $A = a \exp(-iq \cdot x)$ determines the scattering of a neutron by such a field. In the limit of small q it will vary as $qa - aq$ which represents the interaction of a particle possessing a magnetic moment. The first-order interaction between an electron and a neutron is given by the same calculation by considering the exchange of a quantum between the electron and the nucleon. In this case a_μ is q^{-2} times the matrix element of γ_μ between the initial and final states of the electron, the states differing in momentum by q.

The interaction may occur because the neutron of momentum p_1 emits a negative meson becoming a proton which proton interacts with the field and then reabsorbs the meson (Fig. 10a). The matrix for this process is $(p_2 = p_1 + q)$,

$$\int (\gamma_5 k)(p_2 - k - M)^{-1} a(p_1 - k - M)^{-1}(\gamma_5 k)(k^2 - \mu^2)^{-1} d^4 k. \quad (41a)$$

Alternatively it may be the meson which interacts with the field. We assume that it does this in the manner of a scalar potential satisfying the Klein Gordon Eq. (35), (Fig. 10b)

$$-\int (\gamma_5 k_2)(p_1 - k_1 - M)^{-1}(\gamma_5 k_1)(k_2^2 - \mu^2)^{-1}$$
$$\times (k_2 \cdot a + k_1 \cdot a)(k_1^2 - \mu^2)^{-1} d^4 k_1, \quad (42a)$$

where we have put $k_2 = k_1 + q$. The change in sign arises because the virtual meson is negative. Finally there are two terms arising from the $\gamma_5 a$ part of the pseudovector coupling (Figs. 10c, 10d)

$$\int (\gamma_5 k)(p_2 - k - M)^{-1}(\gamma_5 a)(k^2 - \mu^2)^{-1} d^4 k, \quad (43a)$$

and

$$\int (\gamma_5 a)(p_1 - k - M)^{-1}(\gamma_5 k)(k^2 - \mu^2)^{-1} d^4 k. \quad (44a)$$

Using convergence factors in the manner discussed in the section on meson theories each integral can be evaluated and the results combined. Expanded in powers of q the first term gives the magnetic moment of the neutron and is insensitive to the cut-off, the next gives the scattering amplitude of slow electrons on neutrons, and depends logarithmically on the cut-off.

The expressions may be simplified and combined somewhat before integration. This makes the integrals a little easier and also shows the relation to the case of pseudoscalar coupling. For example in (41a) the final $\gamma_5 k$ can be written as $\gamma_5 (k - p_1 + M)$ since $p_1 = M$ when operating on the initial neutron state. This is

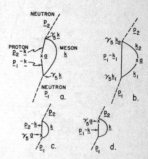

FIG. 10. According to the meson theory a neutron interacts with an electromagnetic potential a by first emitting a virtual charged meson. The figure illustrates the case for a pseudoscalar meson with pseudovector coupling. Appendix D.

$(p_1 - k - M)\gamma_5 + 2M\gamma_5$ since γ_5 anticommutes with p_1 and k. The first term cancels the $(p_1 - k - M)^{-1}$ and gives a term which just cancels (43a). In a like manner the leading factor $\gamma_5 k$ in (41a) is written as $-2M\gamma_5 - \gamma_5(p_2 - k - M)$, the second term leading to a simpler term containing no $(p_2 - k - M)^{-1}$ factor and combining with a similar one from (44a). One simplifies the $\gamma_5 k_1$ and $\gamma_5 k_2$ in (42a) in an analogous way. There finally results terms like (41a), (42a) but with pseudoscalar coupling $2M\gamma_5$ instead of $\gamma_5 k$, no terms like (43a) or (44a) and a remainder, representing the difference in effects of pseudovector and pseudoscalar coupling. The pseudoscalar terms do not depend sensitively on the cut-off, but the difference term depends on it logarithmically. The difference term affects the electron-neutron interaction but not the magnetic moment of the neutron.

Interaction of a proton with an electromagnetic potential can be similarly analyzed. There is an effect of virtual mesons on the electromagnetic properties of the proton even in the case that the mesons are neutral. It is analogous to the radiative corrections to the scattering of electrons due to virtual photons. The sum of the magnetic moments of neutron and proton for charged mesons is the same as the proton moment calculated for the corresponding neutral mesons. In fact it is readily seen by comparing diagrams, that for arbitrary q, the scattering matrix to *first order in the electromagnetic potential* for a proton according to neutral meson theory is equal, if the mesons were charged, to the sum of the matrix for a neutron and the matrix for a proton. This is true, for any type or mixtures of meson coupling, to all orders in the coupling (neglecting the mass difference of neutron and proton).

논문 웹페이지

위대한 논문과의 만남을 마무리하며

이 책은 양자전기역학의 창시자 중 한 명인 파인먼의 1949년 논문에 초점을 맞추었습니다. 또한 이 논문이 나올 수 있게 한 파인먼의 경로 적분에 대한 논문도 다루었습니다.

양자전기역학을 이해하려면 양자장론을 알아야 합니다. 여기서는 양자장론을 일반 독자도 쉽게 이해할 수 있을 만큼만 다루었습니다. 사실 양자장론은 입자 물리 전공 대학원생들이 힘들어하는 과목 중 하나입니다. 그 이유는 엄청나게 많은 수학이 사용되기 때문입니다.

먼저 양자장론의 기초를 다지기 위해 디랙이 도입한 새로운 벡터 기호인 브라켓 기호를 재미있게 소개했습니다. 또한 양자역학이 선형대수라는 수학의 기초 위에 만들어졌으므로 선형대수의 역사를 다루어 보았습니다.

마지막으로 양자전기역학에 대한 파인먼의 논문은 수식보다는 그림으로 이해하는 방향으로 마무리했습니다. 원래의 논문을 이해하려면 복소함수론, 그린 함수 이론과 같은 복잡하고 난해한 수학을 공부해야 하기 때문입니다. 독자의 이해를 돕기 위해 이 시리즈의 《양자혁명》《원자모형》《불확정성원리》《반입자》를 추천합니다. 이 책을 통해 독자들이 양자전기역학의 신비에 푹 빠질 수 있으리라 생각합니다.

출판 기획상 수식을 피할 수 없을 때는 고등학교 수학 정도를 아는 사람이라면 이해할 수 있도록 처음 쓴 원고를 고치고 또 고치는 작업을 반복했습니다. 그렇게 하여 수식을 줄여보려고 했습니다. 하지만 물리를 좋아하는 사람들이 쉽게 따라갈 수 있도록 친절하게 설명했습니다.

원고를 쓰기 위해 20세기의 여러 논문을 뒤적거렸습니다. 지금과는 완연히 다른 용어와 기호 때문에 많이 힘들었습니다. 특히 번역이 안 되어 있는 자료들이 많았지만 프랑스 논문에 대해서는 불문과를 졸업한 아내의 도움으로 조금은 이해할 수 있었습니다.

집필을 끝내자마자 다시 우주팽창에 대한 피블스의 오리지널 논문을 공부하며, 시리즈를 계속 이어나갈 생각을 하니 즐거움에 벅차오릅니다. 제가 느끼는 이 기쁨을 독자들이 공유할 수 있기를 바라며 이제 힘들었지만 재미있었던 양자전기역학에 관한 논문들과의 씨름을 여기서 멈추려고 합니다.

끝으로 용기를 내서 이 시리즈의 출간을 결정한 성림원북스의 이성림 사장과 직원들에게 감사를 드립니다. 시리즈 초안이 나왔을 때,

수식이 많아 출판사들이 꺼릴 것 같다는 생각이 들었습니다. 몇 군데에 출판을 의뢰한 후 거절당하면 블로그에 올릴 생각으로 글을 써 내려갔습니다. 놀랍게도 첫 번째로 이 원고의 이야기를 나눈 성림원북스에서 출간을 결정해 주어서 책이 나올 수 있게 되었습니다. 원고를 쓰는 데 필요한 프랑스 논문의 번역을 도와준 아내에게도 고마움을 전합니다. 그리고 이 책을 쓸 수 있도록 멋진 논문을 만든 고 파인먼 박사님에게도 감사를 드립니다.

진주에서 정완상 교수

이 책을 위해 참고한 논문들

1장

[1] G. Cramer, "Introduction à l'Analyse des lignes Courbes algébriques"(in French), Geneva: Europeana. 656-659. Retrieved 2012—05—18, 1750.

2장

[1] P. A. M. Dirac, A new notation for quantum mechanics, Mathematical Proceedings of the Cambridge Philosophical Society. 35; 416—418, 1939.

[2] P. A. M. Dirac, The Principles of Quantum Mechanics, 1930.

[3] W. Heisenberg, "Über quantentheoretische Umdeutung kinematischer und mechanischer Beziehungen", Zeitschrift für Physik. 33 (1): 879-893, 1925.

[4] M. Born and P. Jordan, "Zur Quantenmechanik", Zeitschrift für Physik. 34 (1): 858-888, 1925.

[5] E. Schrödinger, An Undulatory Theory of the Mechanics of Atoms and Molecules, Phys. Rev. 28; 1049, 1926.

3장

[1] M. S. Vallarta and R. P. Feynman, "The Scattering of Cosmic Rays by the Stars of a Galaxy", Phys. Rev. 55; 506, 1939.

[2] R. P. Feynman, "Forces in Molecules", Physical Review. Vol. 56; 340, 1939.

[3] R. P. Feynman, "Space-Time Approach to Quantum Electrodynamics", Phys. Rev. 76; 769, 1949.

4장

[1] V. Fock, "Konfigurationsraum und zweite Quantelung", Zeitschrift für Physik. 75; 622-647, 1932.

[2] P. Jordan, "Über Verallgemeinerungsmöglichkeiten des Formalismus der Quantenmechanik", Nachr. Akad. Wiss. Göttingen. Math. Phys. Kl. I, 41; 209-217, 1933.

[3] E. P. Wigner, "The unreasonable effectiveness of mathematics in the natural sciences. Richard Courant lecture in mathematical sciences delivered at New York University, May 11, 1959", Communications on Pure and Applied Mathematics. 13 (1): 1-14, 1960.

[4] P. A. M. Dirac, The Quantum Theory of the Emission and Absorption of Radiation, Proceedings of the Royal Society of London. Series A. 114; 243-265, 1927.

수식에 사용하는 그리스 문자

대문자	소문자	읽기	대문자	소문자	읽기
A	α	알파(alpha)	N	ν	뉴(nu)
B	β	베타(beta)	Ξ	ξ	크시(xi)
Γ	γ	감마(gamma)	O	o	오미크론(omicron)
Δ	δ	델타(delta)	Π	π	파이(pi)
E	ε	엡실론(epsilon)	P	ρ	로(rho)
Z	ζ	제타(zeta)	Σ	σ	시그마(sigma)
H	η	에타(eta)	T	τ	타우(tau)
Θ	θ	세타(theta)	Y	υ	입실론(upsilon)
I	ι	요타(iota)	Φ	φ	피(phi)
K	κ	카파(kappa)	X	χ	키(chi)
Λ	λ	람다(lambda)	Ψ	ψ	프시(psi)
M	μ	뮤(mu)	Ω	ω	오메가(omega)

노벨 물리학상 수상자들을 소개합니다

이 책에 언급된 노벨상 수상자는 이름 앞에 ★로 표시하였습니다.

연도	수상자	수상 이유
1901	빌헬름 콘라트 뢴트겐	그의 이름을 딴 놀라운 광선의 발견으로 그가 제공한 특별한 공헌을 인정하여
1902	헨드릭 안톤 로런츠 피터르 제이만	복사 현상에 대한 자기의 영향에 대한 연구를 통해 그들이 제공한 탁월한 공헌을 인정하여
1903	앙투안 앙리 베크렐	자발 방사능 발견으로 그가 제공한 탁월한 공로를 인정하여
	피에르 퀴리 마리 퀴리	앙리 베크렐 교수가 발견한 방사선 현상에 대한 공동 연구를 통해 그들이 제공한 탁월한 공헌을 인정하여
1904	존 윌리엄 스트럿 레일리	가장 중요한 기체의 밀도에 대한 조사와 이러한 연구와 관련하여 아르곤을 발견한 공로
1905	★필리프 레나르트	음극선에 대한 연구
1906	조지프 존 톰슨	기체에 의한 전기 전도에 대한 이론적이고 실험적인 연구의 큰 장점을 인정하여
1907	앨버트 에이브러햄 마이컬슨	광학 정밀 기기와 그 도움으로 수행된 분광 및 도량형 조사
1908	가브리엘 리프만	간섭 현상을 기반으로 사진적으로 색상을 재현하는 방법
1909	굴리엘모 마르코니 카를 페르디난트 브라운	무선 전신 발전에 기여한 공로를 인정받아
1910	요하네스 디데릭 판데르발스	기체와 액체의 상태 방정식에 관한 연구
1911	빌헬름 빈	열복사 법칙에 관한 발견
1912	닐스 구스타프 달렌	등대와 부표를 밝히기 위해 가스 어큐뮬레이터와 함께 사용하기 위한 자동 조절기 발명

1913	헤이커 카메를링 오너스	특히 액체 헬륨 생산으로 이어진 저온에서의 물질 특성에 대한 연구
1914	★막스 폰 라우에	결정에 의한 X선 회절 발견
1915	윌리엄 헨리 브래그 윌리엄 로런스 브래그	X선을 이용한 결정구조 분석에 기여한 공로
1916	수상자 없음	
1917	찰스 글러버 바클라	원소의 특징적인 뢴트겐 복사 발견
1918	★막스 플랑크	에너지 양자 발견으로 물리학 발전에 기여한 공로 인정
1919	★요하네스 슈타르크	커낼선의 도플러 효과와 전기장에서 분광선의 분할 발견
1920	샤를 에두아르 기욤	니켈강 합금의 이상 현상을 발견하여 물리학의 정밀 측정에 기여한 공로를 인정하여
1921	★알베르트 아인슈타인	이론 물리학에 대한 공로, 특히 광전효과 법칙 발견
1922	★닐스 보어	원자 구조와 원자에서 방출되는 방사선 연구에 기여
1923	로버트 앤드루스 밀리컨	전기의 기본 전하와 광전효과에 관한 연구
1924	칼 만네 예오리 시그반	X선 분광학 분야에서의 발견과 연구
1925	제임스 프랑크 구스타프 헤르츠	전자가 원자에 미치는 영향을 지배하는 법칙 발견
1926	장 바티스트 페랭	물질의 불연속 구조에 관한 연구, 특히 침전 평형 발견
1927	아서 콤프턴 ★찰스 톰슨 리스 윌슨	그의 이름을 딴 효과 발견 수증기 응축을 통해 전하를 띤 입자의 경로를 볼 수 있게 만든 방법
1928	오언 윌런스 리처드슨	열전자 현상에 관한 연구, 특히 그의 이름을 딴 법칙 발견
1929	루이 드브로이	전자의 파동성 발견
1930	찬드라세카라 벵카타 라만	빛의 산란에 관한 연구와 그의 이름을 딴 효과 발견
1931	수상자 없음	

1932	★베르너 하이젠베르크	수소의 동소체 형태 발견으로 이어진 양자역학의 창시
1933	★에르빈 슈뢰딩거 ★폴 디랙	원자 이론의 새로운 생산적 형태 발견
1934	수상자 없음	
1935	제임스 채드윅	중성자 발견
1936	빅토르 프란츠 헤스	우주 방사선 발견
	칼 데이비드 앤더슨	양전자 발견
1937	클린턴 조지프 데이비슨 조지 패짓 톰슨	결정에 의한 전자의 회절에 대한 실험적 발견
1938	★엔리코 페르미	중성자 조사에 의해 생성된 새로운 방사성 원소의 존재에 대한 시연 및 이와 관련된 느린중성자에 의한 핵반응 발견
1939	어니스트 로런스	사이클로트론의 발명과 개발, 특히 인공 방사성 원소와 관련하여 얻은 결과
1940 1941 1942	수상자 없음	
1943	오토 슈테른	분자선 방법 개발 및 양성자의 자기 모멘트 발견에 기여
1944	★이지도어 아이작 라비	원자핵의 자기적 특성을 기록하기 위한 공명 방법
1945	★볼프강 파울리	파울리 원리라고도 불리는 배제 원리의 발견
1946	퍼시 윌리엄스 브리지먼	초고압을 발생시키는 장치의 발명과 고압 물리학 분야에서 그가 이룬 발견에 대해
1947	에드워드 빅터 애플턴	대기권 상층부의 물리학 연구, 특히 이른바 애플턴층의 발견
1948	패트릭 메이너드 스튜어트 블래킷	윌슨 구름상자 방법의 개발과 핵물리학 및 우주 방사선 분야에서의 발견
1949	★유카와 히데키	핵력에 관한 이론적 연구를 바탕으로 중간자 존재 예측

연도	수상자	업적
1950	세실 프랭크 파월	핵 과정을 연구하는 사진 방법의 개발과 이 방법으로 만들어진 중간자에 관한 발견
1951	존 더글러스 콕크로프트 어니스트 토머스 신턴 월턴	인위적으로 가속된 원자 입자에 의한 원자핵 변환에 대한 선구자적 연구
1952	펠릭스 블로흐 에드워드 밀스 퍼셀	핵자기 정밀 측정을 위한 새로운 방법 개발 및 이와 관련된 발견
1953	프리츠 제르니커	위상차 방법 시연, 특히 위상차 현미경 발명
1954	★막스 보른 발터 보테	양자역학의 기초 연구, 특히 파동함수의 통계적 해석 우연의 일치 방법과 그 방법으로 이루어진 그의 발견
1955	윌리스 유진 램 폴리카프 쿠시	수소 스펙트럼의 미세 구조에 관한 발견 전자의 자기 모멘트를 정밀하게 측정한 공로
1956	윌리엄 브래드퍼드 쇼클리 존 바딘 월터 하우저 브래튼	반도체 연구 및 트랜지스터 효과 발견
1957	양전닝 리정다오	소립자에 관한 중요한 발견으로 이어진 소위 패리티 법칙에 대한 철저한 조사
1958	파벨 알렉세예비치 체렌코프 일리야 프란크 이고리 탐	체렌코프 효과의 발견과 해석
1959	에밀리오 지노 세그레 오언 체임벌린	반양성자 발견
1960	도널드 아서 글레이저	거품 상자의 발명
1961	로버트 호프스태터	원자핵의 전자 산란에 대한 선구적인 연구와 핵자 구조에 관한 발견
	루돌프 뫼스바워	감마선의 공명 흡수에 관한 연구와 그의 이름을 딴 효과에 대한 발견

연도	수상자	공적
1962	★레프 다비도비치 란다우	응집 물질, 특히 액체 헬륨에 대한 선구적인 이론
1963	★유진 폴 위그너	원자핵 및 소립자 이론에 대한 공헌, 특히 기본 대칭 원리의 발견 및 적용을 통한 공로
	마리아 괴페르트 메이어	핵 껍질 구조에 관한 발견
	한스 옌센	
1964	니콜라이 바소프	메이저-레이저 원리에 기반한 발진기 및 증폭기의 구성으로 이어진 양자 전자 분야의 기초 작업
	알렉산드르 프로호로프	
	찰스 하드 타운스	
1965	★도모나가 신이치로	소립자의 물리학에 심층적인 결과를 가져온 양자전기역학의 근본적인 연구
	★줄리언 슈윙거	
	★리처드 필립스 파인먼	
1966	알프레드 카스틀레르	원자에서 헤르츠 공명을 연구하기 위한 광학적 방법의 발견 및 개발
1967	★한스 알브레히트 베테	핵반응 이론, 특히 별의 에너지 생산에 관한 발견에 기여
1968	루이스 월터 앨버레즈	소립자 물리학에 대한 결정적인 공헌, 특히 수소 기포 챔버 사용 기술 개발과 데이터 분석을 통해 가능해진 다수의 공명 상태 발견
1969	★머리 겔만	기본 입자의 분류와 그 상호 작용에 관한 공헌 및 발견
1970	한네스 올로프 예스타 알벤	플라스마 물리학의 다양한 부분에서 유익한 응용을 통해 자기유체역학의 기초 연구 및 발견
	루이 외젠 펠릭스 네엘	고체물리학에서 중요한 응용을 이끈 반강자성 및 강자성에 관한 기초 연구 및 발견
1971	데니스 가보르	홀로그램 방법의 발명 및 개발
1972	존 바딘	일반적으로 BCS 이론이라고 하는 초전도 이론을 공동으로 개발한 공로
	리언 닐 쿠퍼	
	존 로버트 슈리퍼	

1973	에사키 레오나	반도체와 초전도체의 터널링 현상에 관한 실험적 발견
	이바르 예베르	
	브라이언 데이비드 조지프슨	터널 장벽을 통과하는 초전류 특성, 특히 일반적으로 조지프슨 효과로 알려진 현상에 대한 이론적 예측
1974	마틴 라일	전파 천체물리학의 선구적인 연구: 라일은 특히 개구 합성 기술의 관찰과 발명, 그리고 휴이시는 펄서 발견에 결정적인 역할을 함
	앤터니 휴이시	
1975	오게 닐스 보어	원자핵에서 집단 운동과 입자 운동 사이의 연관성 발견과 이 연관성에 기초한 원자핵 구조 이론 개발
	★벤 로위 모텔손	
	제임스 레인워터	
1976	버턴 릭터	새로운 종류의 무거운 기본 입자 발견에 대한 선구적인 작업
	새뮤얼 차오 충 팅	
1977	필립 워런 앤더슨	자기 및 무질서 시스템의 전자 구조에 대한 근본적인 이론적 조사
	네빌 프랜시스 모트	
	존 해즈브룩 밴블렉	
1978	표트르 레오니도비치 카피차	저온 물리학 분야의 기본 발명 및 발견
	아노 앨런 펜지어스	우주 마이크로파 배경 복사의 발견
	로버트 우드로 윌슨	
1979	★셸던 리 글래쇼	특히 약한 중성 전류의 예측을 포함하여 기본 입자 사이의 통일된 약한 전자기 상호 작용 이론에 대한 공헌
	압두스 살람	
	스티븐 와인버그	
1980	제임스 왓슨 크로닌	중성 K 중간자의 붕괴에서 기본 대칭 원리 위반 발견
	밸 로그즈던 피치	
1981	니콜라스 블룸베르헌	레이저 분광기 개발에 기여
	아서 레너드 숄로	
	카이 만네 뵈리에 시그반	고해상도 전자 분광기 개발에 기여

연도	수상자	업적
1982	케네스 게디스 윌슨	상전이와 관련된 임계 현상에 대한 이론
1983	수브라마니안 찬드라세카르	별의 구조와 진화에 중요한 물리적 과정에 대한 이론적 연구
	윌리엄 앨프리드 파울러	우주의 화학 원소 형성에 중요한 핵반응에 대한 이론 및 실험적 연구
1984	카를로 루비아	약한 상호 작용의 커뮤니케이터인 필드 입자 W와 Z의 발견으로 이어진 대규모 프로젝트에 결정적인 기여
	시몬 판데르 메이르	
1985	클라우스 폰 클리칭	양자화된 홀 효과의 발견
1986	에른스트 루스카	전자 광학의 기초 작업과 최초의 전자 현미경 설계
	게르트 비니히	스캐닝 터널링 현미경 설계
	하인리히 로러	
1987	요하네스 게오르크 베드노르츠	세라믹 재료의 초전도성 발견에서 중요한 돌파구
	카를 알렉산더 뮐러	
1988	리언 레더먼	뉴트리노 빔 방법과 뮤온 중성미자 발견을 통한 경입자의 이중 구조 증명
	멜빈 슈워츠	
	잭 스타인버거	
1989	노먼 포스터 램지	분리된 진동 필드 방법의 발명과 수소 메이저 및 기타 원자시계에서의 사용
	한스 게오르크 데멜트	이온 트랩 기술 개발
	볼프강 파울	
1990	제롬 프리드먼	입자 물리학에서 쿼크 모델 개발에 매우 중요한 역할을 한 양성자 및 구속된 중성자에 대한 전자의 심층 비탄성 산란에 관한 선구적인 연구
	헨리 웨이 켄들	
	리처드 테일러	
1991	피에르질 드 젠	간단한 시스템에서 질서 현상을 연구하기 위해 개발된 방법을 보다 복잡한 형태의 물질, 특히 액정과 고분자로 일반화할 수 있음을 발견

1992	조르주 샤르파크	입자 탐지기, 특히 다중 와이어 비례 챔버의 발명 및 개발
1993	러셀 헐스	새로운 유형의 펄서 발견, 중력 연구의 새로운 가능성을 연 발견
	조지프 테일러	
1994	버트럼 브록하우스	중성자 분광기 개발
	클리퍼드 셜	중성자 회절 기술 개발
1995	마틴 펄	타우 렙톤의 발견
	프레더릭 라이너스	중성미자 검출
1996	데이비드 리	헬륨-3의 초유동성 발견
	더글러스 오셔로프	
	로버트 리처드슨	
1997	스티븐 추	레이저 광으로 원자를 냉각하고 가두는 방법 개발
	클로드 코엔타누지	
	윌리엄 필립스	
1998	로버트 로플린	부분적으로 전하를 띤 새로운 형태의 양자 유체 발견
	호르스트 슈퇴르머	
	대니얼 추이	
1999	헤라르뒤스 엇호프트	물리학에서 전기약력 상호작용의 양자 구조 규명
	마르티뉴스 펠트만	
2000	조레스 알표로프	정보 통신 기술에 대한 기초 작업(고속 및 광전자 공학에 사용되는 반도체 이종 구조 개발)
	허버트 크로머	
	잭 킬비	정보 통신 기술에 대한 기초 작업(집적회로 발명에 기여)
2001	에릭 코넬	알칼리 원자의 희석 가스에서 보스-아인슈타인 응축 달성 및 응축 특성에 대한 초기 기초 연구
	칼 위먼	
	볼프강 케테를레	

연도	수상자	업적
2002	레이먼드 데이비스	천체물리학, 특히 우주 중성미자 검출에 대한 선구적인 공헌
	고시바 마사토시	
	리카르도 자코니	우주 X선 소스의 발견으로 이어진 천체물리학에 대한 선구적인 공헌
2003	알렉세이 아브리코소프	초전도체 및 초유체 이론에 대한 선구적인 공헌
	비탈리 긴즈부르크	
	앤서니 레깃	
2004	데이비드 그로스	강한 상호작용 이론에서 점근적 자유의 발견
	데이비드 폴리처	
	프랭크 윌첵	
2005	★로이 글라우버	광학 일관성의 양자 이론에 기여
	존 홀	광 주파수 콤 기술을 포함한 레이저 기반 정밀 분광기 개발에 기여
	테오도어 헨슈	
2006	존 매더	우주 마이크로파 배경 복사의 흑체 형태와 이방성 발견
	조지 스무트	
2007	알베르 페르	자이언트 자기 저항의 발견
	페터 그륀베르크	
2008	난부 요이치로	아원자 물리학에서 자발적인 대칭 깨짐 메커니즘 발견
	고바야시 마코토	자연계에 적어도 세 종류의 쿼크가 존재함을 예측하는 깨진 대칭의 기원 발견
	마스카와 도시히데	
2009	찰스 가오	광 통신을 위한 섬유의 빛 전송에 관한 획기적인 업적
	윌러드 보일	영상 반도체 회로(CCD 센서)의 발명
	조지 엘우드 스미스	
2010	안드레 가임	2차원 물질 그래핀에 관한 획기적인 실험
	콘스탄틴 노보셀로프	

2011	솔 펄머터 브라이언 슈밋 애덤 리스	원거리 초신성 관측을 통한 우주 가속 팽창 발견
2012	세르주 아로슈 데이비드 와인랜드	개별 양자 시스템의 측정 및 조작을 가능하게 하는 획기적인 실험 방법
2013	프랑수아 앙글레르 피터 힉스	아원자 입자의 질량 기원에 대한 이해에 기여하고 최근 CERN의 대형 하드론 충돌기에서 ATLAS 및 CMS 실험을 통해 예측된 기본 입자의 발견을 통해 확인된 메커니즘의 이론적 발견
2014	아카사키 이사무 아마노 히로시 나카무라 슈지	밝고 에너지 절약형 백색 광원을 가능하게 한 효율적인 청색 발광 다이오드의 발명
2015	가지타 다카아키 아서 맥도널드	중성미자가 질량을 가지고 있음을 보여주는 중성미자 진동 발견
2016	데이비드 사울레스 덩컨 홀데인 마이클 코스털리츠	위상학적 상전이와 물질의 위상학적 위상에 대한 이론적 발견
2017	라이너 바이스 킵 손 배리 배리시	LIGO 탐지기와 중력파 관찰에 결정적인 기여
2018	아서 애슈킨	레이저 물리학 분야의 획기적인 발명(광학 핀셋과 생물학적 시스템에 대한 응용)
	제라르 무루 도나 스트리클런드	레이저 물리학 분야의 획기적인 발명(고강도 초단파 광 펄스 생성 방법)
2019	제임스 피블스	우주의 진화와 우주에서 지구의 위치에 대한 이해에 기여(물리 우주론의 이론적 발견)
	미셸 마요르 디디에 쿠엘로	우주의 진화와 우주에서 지구의 위치에 대한 이해에 기여(태양형 항성 주위를 공전하는 외계 행성 발견)

연도	수상자	업적
2020	로저 펜로즈	블랙홀 형성이 일반 상대성 이론의 확고한 예측이라는 발견
	라인하르트 겐첼	우리 은하의 중심에 있는 초거대 밀도 물체 발견
	앤드리아 게즈	
2021	마나베 슈쿠로	복잡한 시스템에 대한 이해에 획기적인 기여(지구 기후의 물리적 모델링, 가변성을 정량화하고 지구 온난화를 안정적으로 예측)
	클라우스 하셀만	
	조르조 파리시	복잡한 시스템에 대한 이해에 획기적인 기여 (원자에서 행성 규모에 이르는 물리적 시스템의 무질서와 요동의 상호작용 발견)
2022	알랭 아스페	얽힌 광자를 사용한 실험, 벨 불평등 위반 규명 및 양자 정보 과학 개척
	존 클라우저	
	안톤 차일링거	
2023	피에르 아고스티니	물질의 전자 역학 연구를 위해 아토초(100경분의 1초) 빛 펄스를 생성하는 실험 방법 고안
	페렌츠 크라우스	
	안 륄리에	
2024	존 홉필드	인공신경망을 이용해 머신러닝을 가능하게 하는 기초적인 발견과 발명
	제프리 힌턴